Contents

DEPARTMENT OF THE ENVIRONMENT

Slate Waste Tips and Workings in Britain

Richards, Moorehead & Laing Ltd.

LONDON: HMSO

ISBN 0 11 752971 0

Cover photograph: slate extraction and processing
since 1816 have produced this quarrying landscape at
Pen yr Orsedd, Nantlle, Gwynedd.

PHOTOGRAPH COURTESY OF GWYNEDD COUNTY COUNCIL

Appendices

References

List of Figures

List of Tables

Photographs

List of Plates

Acknowledgements

This report has been prepared by Richards, Moorehead and Laing Ltd. Engineering, Environmental and Landscape Consultants.

The contractor's team was:

Mr N Ward	(Project Director)	Dr I Brown (Independent Consultant)
Mr S Blunt	(Nominated Officer)	Mr J Riden (Independent Consultant)
Mr P Hay		Mr M Williams (Independent Consultant)
Mr B Houldsworth		
Dr J Palmer		
Mr I Richards		

The research was commissioned by the Department of the Environment (Minerals Division). The contractors are particularly grateful for the guidance and encouragement given by:

Dr Tom Simpson   (DOE, Nominated Officer)

Project Steering Committee:

Miss A Ward, Chair	(DOE)
Dr M Scott	(DOE)
Mr B Courtier	(Welsh Office)
Mr N Fecitt	(Natural Slate Quarries Association/Kirkstone Quarries Ltd)
Mr J Gibbins	(Association of County Councils/County Planning Officers Society)
Mr P Gordon	(Cumbria County Council)
Mr T Hughes	(Natural Slate Quarries Association/A McAlpine Slate Products Ltd) succeeded by Mr C Law
Mr R Lawday	(Welsh Development Agency)
Mr P Matcham	(Lake District National Park Authority)
Mr C Morgan	(Welsh Office)
Mr R Toms	(Natural Slate Quarries Association)
Mr R Sykes, Secretary	(DOE)

The contractors also wish to acknowledge the invaluable assistance provided by the slate producers, regulatory authorities and many other organisations and individuals consulted during the preparation of this report.

Executive Summary

Background

Slate has been worked in Britain for many centuries, providing roofing and building materials, but also generating vast quantities of waste. It is estimated that there are some 500,000,000 tonnes of slate waste in tips in the main areas of active and former production, which are Cornwall, Cumbria, North and Mid Wales, and the Highlands of Scotland. Some of this material is suitable for use as a secondary aggregate or raw material, but its removal on a large scale would cause significant disturbance in many cases. Approximately 300 ha of former slate workings have been treated in reclamation schemes, but 600 ha still lies derelict and 1500 ha could become derelict when slate working ceases. The restoration standards expected of mineral operators are increasing. The Government has set out proposals to update operating and restoration conditions at existing, older sites. There is therefore a growing need for guidance on management and treatment of land affected by working of slate.

To meet this need, the Department of the Environment commissioned Richards, Moorehead and Laing Ltd to examine the need for reclaiming slate waste tips and quarries in Britain. The objective of the study was to provide technical guidance on the approaches and methods available to rehabilitate workings and *in-situ* waste tips, and to examine the environmental consequences of removing slate waste for use elsewhere. During the research programme the study team gathered published and unpublished literature, visited many sites throughout Britain, and consulted a wide range of organisations including:

- quarry operators;
- local government;
- statutory bodies;
- non-statutory organisations;
- voluntary groups and individuals.

The team also developed and tested a method for the assessment of land use potential and reclamation options, using Blaenau Ffestiniog in North Wales as a pilot study area. The findings of the research and pilot study have been distilled to provide guidance on the assessment of slate working sites and restoration or treatment options, including the selection of new uses for these sites, and the techniques which are available to prepare sites for these uses.

Description of slate workings

The majority of current slate-producing sites in Britain commenced in the 18th or 19th centuries, and are now extracting from or tipping on land with little or no requirements in the planning permission for restoration. The smallest workings employ just a few men, extracting slate from a single face, although these are often in the midst of large, abandoned workings. The largest active open-pits cover up to a hundred hectares and employ over a hundred people. Underground working continues in Gwynedd, although two centuries of activity have left several hundred hectares of waste tips on the surface. As slate extraction rarely, if ever, proceeds systematically through a deposit until the reserve is exhausted, slate quarries can remain operational for many decades or centuries with no prospect of restoration. Slate waste tips are often considered by their owners to be an asset for future use. This consideration may be an obstacle to reclamation or new uses.

Abandoned slate workings can present hazards to casual visitors. These hazards include unguarded quarry faces, deep flooded excavations and deteriorating structures. Slate waste tips are inherently stable unless disturbed, and no instances of spontaneous tip movement were found during this research. Slate waste is coarse-grained and free-draining, and tips are unaffected by through passage of water.

At many sites, slate waste tips are retained by unmortared slate walls which deteriorate progressively. Such tips have caused concern where collapse would endanger the public and so a number of reclamation schemes have been carried out to strengthen these retaining walls or to regrade the retained tips.

The historical remains of slate workings are a source of great interest to industrial archaeologists, and are of growing interest to the general public. A number of sites have been prepared as open-air or underground visitor centres. Many more sites contain examples of water power, of transport systems, processing buildings and the like, some of which can be seen from public footpaths. Sites with public access can, with sensitive consolidation and low-key interpretation, add greatly to public enjoyment of the countryside.

Slate waste tips are generally much more resistant to natural colonisation by vegetation than are most other forms of mineral extraction waste. This is due to the coarse particle size and extremely free drainage of slate waste, its resistance to weathering and lack of available plant nutrients. Many sites are elevated and exposed, further limiting the rate of colonisation. Abandoned slate workings and associated buildings do provide breeding and roosting sites to birds, bats and other mammals, and undisturbed land within workings sometimes supports semi-natural vegetation. Some old slate workings have therefore contributed to the diversity of wildlife habitats. Quarrying or the deposition of spoil inevitably causes the loss of the habitat which is present, and the development of new habitats will be very slow unless the land is restored and managed appropriately.

The slate workings of Britain are all predominately rural areas of high landscape quality. In many areas, quarrying was the most powerful human influence leading to the present landscape. Perceptions of slate workings and especially the waste tips, range from an unacceptable disfigurement of otherwise beautiful landscapes, to an essential component of the cultural heritage of communities which only exist because of the slate industry. The management or treatment of slate workings should involve finding a balance between these perceptions, and seeking to reduce the negative impacts of workings without reducing their historical interest.

The use of slate waste

Many attempts have been made to find new uses for slate waste, but with the exception of those noted here, economic or practical considerations have prevented successful exploitation of the waste tips. Slate waste is a varied material, and so not all wastes are suitable for the current uses, which include:

- tiles, bitumastic coatings, face powder and many other products which use powdered slate as an inert constituent;
- roofing felt, which is coated with slate granules of various colours;
- construction aggregates, including DOT Type 1 and Type 2 sub-base for road construction;
- walling and paving stone, often salvaged from waste tips;
- bulk fill material for landfilling, embankment construction and similar applications.

The use of slate waste on a large scale as a 'secondary' mineral is generally constrained by the cost of transporting it to the point of use. Few slate workings are close to rail lines or sea ports and none use these modes of long-distance transport. 'Primary' aggregate sources closer to areas of demand can supply materials in bulk more cheaply than the producers of slate-derived products. The total consumption of slate waste is insignificant in relation to the volume of waste being generated or the existing waste tips.

An examination of the environmental consequences of large-scale slate waste removal showed that site-specific considerations are likely to be more significant than general comparisons between primary and secondary sources, although the use of wastes arising from current slate production would cause very little disturbance at the site of production and could provide substantial benefits. Many abandoned slate workings have poor road access, significant natural colonisation, remains of industrial archaeological interest, and low-key recreational uses which would all be disturbed by the removal of slate waste. The best environmental practice would be to concentrate the large-scale use/re-use of slate wastes at the sites of current arisings and from waste tips where the environmental problems are minimal or outweighed by the benefits of slate waste removal. Selective, small-scale waste removal for local use can reduce the need to transport materials from elsewhere and can form part of a reclamation scheme to prepare land for new uses.

Active workings in Blaenau Ffestiniog and Bethesda, Gwynedd and at Kirkby in Furness, Cumbria, are close to rail lines or seaports. With some financial support, several million tonnes of aggregates could be produced each year and transported to contribute to national demands for aggregates.

New uses of abandoned slate workings

Many abandoned slate workings are now put to new uses by the public as individuals or in groups, often without any site preparation or even the consent of the owner. These uses include:

- walking and informal recreation,
- study of industrial archaeology, wildlife, geology;
- outdoor sports such as climbing, abseiling and sub-aqua.

Some sites have been occupied by travellers. Others now contain abandoned cars and fly-tipped wastes.

Informal and passive uses such as woodland and grazing were found during this research. Formal site uses are likely to require some works to prepare the site and to reduce the hazards to site users. Works constituting development or engineering operations will require planning permission. Sites have been developed as industrial estates and for housing, tourism centres, roads and car parks. Permission has been granted for inert waste disposal at some sites, but the disposal of domestic refuse is unlikely to be permitted since it is difficult to meet requirements to contain and treat leachates.

Methods of reclamation

In the period 1972 - 1992, sixteen land reclamation schemes were carried out at derelict slate workings with the aid of land reclamation grants. All these schemes largely or completely met the objectives that were set for them, although further work would improve the integration of these sites with the surrounding landscapes.
The principal elements were:

- clearance of structures and debris;
- diversion or removal of water;
- excavation and repositioning of slate waste;
- compaction of fill;
- trimming of regraded surfaces;
- surface crushing, preparation and seeding;
- provision of surface water drainage;
- planting, fencing and landscape maintenance;
- provision of infrastructure for new uses.

In most of the schemes, large volumes of slate waste were moved, using a blend of civil engineering and quarrying techniques to deal with the extremely coarse material. Slate was placed in layers and spread using bulldozers. The surface layers were compacted with rollers where development was to take place. These methods have prevented settlement of the fill. In all the schemes, provision was made to collect and discharge the surface water run-off, and the drainage water which can issue from the base of the slate waste tips during prolonged heavy rainfall. In the absence of topsoil, large areas were prepared for grass seeding by crushing the surface with heavy grid rollers. This produced a thin layer of finer slate particles in which grass would establish if nitrogen and phosphate were supplied. These swards proved vulnerable to uncontrolled grazing and to an absence of regular fertiliser applications. Regression of these vulnerable swards illustrates the need to match the land use objectives and the selection of species with the resources available for management. Trees have successfully been established in beds of soil, in layers of finely-crushed slate, and in pockets of compost. Species which naturally colonise slate waste tips, such as Birch, Rowan and Goat Willow, have been found to perform well due to their tolerance of drought, lack of nutrients and exposure. If nitrogen-fixing trees, shrubs and legumes can be maintained within woodlands or grass swards, the growth of all species will be improved. A reclamation project can only be fully successful if the vegetation which is established is sustainable under the conditions of use and maintenance which follow the initial scheme.

Assessing land use potential and reclamation options

Slate working sites have many factors, positive and negative, which should be considered during the selection of new site uses and in preparing updated schemes for rehabilitation under existing permissions. Sites can meet some of the land use demands of an area if appropriate modifications are made. A framework for site assessment is set out, which aims to assist the user to match the potential sites with requirements of the land uses which are in demand. The framework includes an assessment of the feasibility of slate waste removal, so that land uses may be considered for the site both in its undisturbed and reshaped condition.

Site assessment is a multi-disciplinary process involving:
- the stability of tips, quarries and structures;
- potential for use of slate waste;
- the accessibility of the site;
- drainage;
- wildlife value;
- landscape assessment;
- geological interest;
- recreation potential;
- industrial archaeological value;
- redevelopment potential;

An approach to the systematic assessment of slate-working landscapes was developed as part of the research and is presented in this report.

Case studies

The report contains seven case studies drawn from North Wales, the Lake District and Scotland. These case studies illustrate issues such as:
- planning control and site restoration;
- conflicts with wildlife and valued landscapes;
- sources of secondary minerals, and the environmental consequences of extraction;
- tourism, industrial archaeology and conservation;
- waste disposal;
- reclamation and long-term management;
- informal uses, including woodland, walking, climbing and abseiling.

Crynodeb Gweithredol

Cefndir

Bu cloddio am lechi yn digwydd ym Mhrydain ers canrifoedd lawer, gan ddarparu defnyddiau toi ac adeiladu, ond hefyd yn cynhyrchu tomenni enfawr o wastraff. Amcangyfrifir bod tua 500,000,000 o dunelli o wastraf llechi mewn tomenni yn y prif ardaloedd fu'n cynhyrchu llechi, neu sydd yn dal i wneud hynny, sef Cernyw, Cumbria, Gogledd a Chanolbarth Cymru ac Ucheldir yr Alban. Mae rhyw gymaint o'r deunydd hwn yn addas i'w ddefnyddio fel cerrig mâl eilradd neu ddeunydd crau, ond byddai ei glirio ar raddfa fawr yn achosi aflonyddwch sylweddol mewn llawer achos. Eisoes cafodd rhyw 300 ha o gyn-weithfeydd llechi eu trin mewn cynlluniau adfer, ond mae rhyw 600 ha yn dal yn dir diffaith, a gallai 1500 ha fynd yn ddiffaith wrth i waith llechi ddod i ben. Mae'r safonau adfer a ddisgwylir gan weithredwyr mwynau yn codi. Mae'r Llywodraeth wedi gosod allan gynigion i ddiweddaru amodau gweithio ac adfer ar y safleoedd presennol, hyn. Felly mae angen cynyddol am arweiniad ar reoli a thrin tir yr effeithiwyd arno gan weithfeydd llechi.

Er mwyn cwrdd â'r angen hwn, mae Adran yr Amgylchedd wedi comisiynu Richards, Moorhead a Laing Cyf i archwilio'r angen am adfer tomenni gwastraff a chwareli llechi ym Mhrydain. Amcan yr astudiaeth oedd darparu arweiniad technegol ar yr ymagwedd a'r dulliau sydd ar gael i ymaddasu gweithfeydd a thomenni gwastraff yn y fan a'r lle, ac i archwilio canlyniadau amgylcheddol clirio'r gwastraff llechi i'w ddefnyddio rhywle arall. Yn ystod y rhaglen ymchwil casglodd y tim astudio lenyddiaeth - cyhoeddedig ac heb ei gyhoeddi - ymwelwyd ag amryw o safleoedd ar hyd a lled Prydain, ac ymgynghorwyd ag amrediad eang o sefydliadau, yn cynnwys:
- gweithredwyr chwareli;
- llywodraeth leol;
- cyrff statudol;
- sefydliadau anstatudol;
- grwpiau gwirfoddol ac unigolion.

Hefyd fe wnaeth y tim ddatblygu ac arbrofi dull o asesu defnydd potensial ar dir a dewisiadau adfer, gan ddefnyddio Blaenau Ffestiniog yng Ngogledd Cymru fel ardal astudiaeth beilot. Distyllwyd canlyniadau'r ymchwil a'r astudiaeth beilot er mwyn cynnig arweiniad ar asesu safleoedd gweithio llechi ac ar y dewisiadau adfer neu driniaeth, gan gynnwys defnydd newydd i'r safleoedd hyn, a'r technegau sydd ar gael i baratoi safleoedd ar gyfer y defnydd newydd arnynt.

Disgrifiad o weithfeydd llechi

Dechreuodd y rhan fwyaf o'r safleoedd cynhyrchu sydd ym Mhrydain yn ystod y 18ed a'r 19eg Ganrif, ac maent erbyn hyn yn cloddio tir, neu daflu tomenni arno, heb fawr neu ddim gofynion yn y caniatad cynllunio i adfer y tir. Mae'r gweithfeydd lleiaf yn cyflogi ond ychydig

ddynion, yn tynnu llechi o un wyneb, er bod rhain yn aml yng nghanol gweithfeydd enfawr a adawyd. Mae'r ceudyllau agored mwyaf sydd yn dal i weithio yn cynnwys hyd at gan hectar o dir ac yn cyflogi dros gant o bobl. Mae gweithio dan y ddaear yn parhau yng Ngwynedd, er bod dwy ganrif o weithgarwch wedi gadael rhai cannoedd o hectarau o domenni gwastraff ar y wyneb. Gan nad yw cloddio llechi byth, os o gwbl, yn mynd yn ei flaen yn systmatig drwy'r graig nes bod y ffynhonnell wedi darfod, gall chwareli llechi fod yn weithredol am ddegau oflynyddoedd neu ganrifoedd, heb fawr o obaith eu hadfer. Yn aml iawn mae eu perchnogion yn edrych ar domenni llechi fel ased i'w defnyddio yn y dyfodol. Gallai agwedd felly fod yn rhwystr i'w hadfer neu i wneud defnydd newydd ohonynt.

Gall gweithfeydd llechi a adawyd fod yn beryglus i ymwelwyr achlysurol. Gall y peryglon gynnwys wynebau chwarel heb eu gwarchod, ceudyllau dwfn wedi gorlifo ac adeiladau sy'n dirywio. Yn eu hanfod mae tomenni gwastraff llechi yn gadarn, os na chânt eu haflonyddu, ac ni ddaethpwyd ar draws unrhyw enghraifft o domen yn symud yn ystod y gwaith ymchwil hwn. Mae graen bras i wastraff llechi, ac mae'n draenio'n rhwydd, ac nid yw dwr yn llifo drwyddynt yn effeithio ar domenni.

Mewn llawer safle, waliau llechi sychion sydd yn cynnal y tomenni gwastraff, ac mae rheiny yn dirywio mwy a mwy. Mae tomenni felly wedi bod yn achos pryder lle gallent fod yn beryglus i'r cyhoedd pe baent yn llithro, ac felly gwnaed nifer o gynlluniau adfer i gryfhau'r waliau cynnal hyn neu i ail raddio'r tomenni a gynhelir ganddynt.

Mae olion hanesyddol gweithfeydd llechi o ddiddordeb mawr i archaeolegwyr diwydiannol, ac o ddiddordeb cynyddol i'r cyhoedd yn gyffredinol. Mae nifer o safleoedd wedi'u datblygu fel canolfannau ymwelwyr, rhai awyr agored a rhai dan y ddaear. Mae llawer mwy o safleoedd yn cynnwys enghreifftiau o ynni dwr, sustemau cludiant, adeiladau prosesu ac yn y blaen, rhai ohonynt yn weladwy o lwybrau cyhoeddus. Gydag ychydig o waith atgyfnerthu a dehongli ar raddfa fechan, gallai safleoedd sydd â mynediad cyhoeddus ychwanegu'n fawr at fwynhad y cyhoedd o gefn gwlad.

Fel arfer mae tomenni gwastraff llechi yn llawer mwy gwrthwynebus i lystyfiant wladychu'n naturiol arnynt nag ydyw mathau eraill o wastraff cloddio mwynau. Mae hyn am fod y gronynnau'n fwy bras ac yn draenio'n rhyfeddol o rwydd, am nad yw'r tywydd yn effeithio fawr ddim arno ac nad oes maethyddion i blanhigion ynddo. Mae llawer o'r safleoedd ar dir uchel ac yn agored i'r tywydd, gan fod yn rhwystr pellach i wladychu arnynt. Mae gweithfeydd llechi a adawyd a'u hadeiladau perthynol yn fagwrfa a nythle i adar, ystlumod a mamaliaid eraill, ac mae tir na chafodd ei aflonyddu mewn gweithfeydd weithiau'n cynnal llystyfiant lled-naturiol. Felly mae rhai hen weithfeydd llechi wedi cyfrannu at amrywiaeth cynefinoedd bywyd gwyllt. Yn anorfod, mae gwaith cloddio neu daflu gwastraff yn achosi colli cynefinoedd sydd yno'n barod, ac araf iawn fydd cynefinoedd newydd yn datblgyu oni chaiff y tir ei adfer a'i reoli'n iawn.

Mae'r rhan fwyaf o weithfeydd llechi Prydain mewn ardaloedd gwledig, o ansawdd uchel o ran eu golygfeydd. Mewn llawer ardal, chwarelydda oedd y dylanwad mwyaf pwerus a greodd y tirwedd presennol. Mae'r farn ar weithfeydd llechi, ac yn arbennig ar domenni gwastraff, yn amrywio o hagru'n annerbyniol tirwedd a oedd yn hardd, i feddwl amdanynt fel rhan hanfodol o dreftadaeth ddiwylliannol cymunedau na fyddai'n bod oni bai am y diwydiant llechi. Dylai rheoli neu drin gweithfeydd llechi olygu cael cydbwysed rhwng y ddwy farn, a cheisio lliniaru effeithiau negyddol gweithfeydd, heb amharu ar eu diddordeb hanesyddol.

Defnyddio gwastraff llechi

Gwnaed llawer ymdrech i ddod o hyd i ffyrddnewydd o ddefnyddio gwatraff llechi, ond ac eithrio'r rhai a nodir yma, mae ystyriaethau economaidd neu rai ymarferol wedi rhwystro unrhyw ymelwa llwyddiannus ar domenni llechi. Mae gwastraff llechi yn ddefnydd amrywiol, ac felly nid yw pob gwastraff yn addas ar gyfer y defnydd cyfredol, sydd yn cynnwys:
- llechi toi, arwisg fitiwmen, powdr wyneb, a nifer fawr o gynhyrchion eraill sydd yn defnyddio llechi mâl fel cyfansoddyn diegni;
- ffelt toi, a arwisgir â gronynau llechi mewn amryw o liwiau;
- cerrig mâl adeiladu, yn cynnwys is-sylfaen Math 1 a Math 2 Yr Adran Drafnidiaeth ar gyfer adeiladu ffyrdd;
- cerrig waliau a phalmentydd, yn aml wedi'u cymryd yn syth o'r tomenni;
- defnydd llanw ar gyfer mewnlenwi tir, adeiladu argloddiau a gwaith tebyg.

Ar y cyfan cyfyngir ar ddefnyddio gwastraff llechi ar raddfa fawr fel mwyn "eilradd" gan y gost o'i gludo i'r man lle defnyddid ef. Ychydig iawn o weithfeydd llechi sydd yn ymyl rheilffordd neu borthladdoedd, ac nid oes yr un yn defnyddio'r mathau hyn o gludo. Mae ffynonellau cerrig mâl "cynradd" yn nes at yr ardaloedd lle mae galw amdano yn medru cyflewni crynswth yn rhatach na chynhyrchwyr deunydd sy'n deillio o lechi. Nid yw'r defnydd a wneir ar wastraff llechi yn fawr ddim o'i gymharu â faint o wastraff a gynhyrchir, neu'r tomenni sy'n bod eisoes.

Mae archwiliad ar ganlyniadau amgylcheddol clirio gwastraff llechi ar raddfa fawr wedi dangos y byddai ystyriaethau penodol ar y safle yn fwy arwyddocaol na chymhariaeth gyffredinol rhwngffynonellau cynradd ac eilradd, er na fyddai defnyddio gwastraff sy'n deillio o'r gwaith cynhyrchu presennol yn achosi fawr o aflonyddwch yn y man lle cynhyrchir ef, a gallai gynnig manteision sylweddol. Ffyrdd mynediad drwg sydd i lawer o'r gweithfeydd llechi a adawyd, a byddai clirio gwastraff oddi arnynt yn aflonyddu ar lystyfiant naturiol sylweddol, olion sydd o ddiddordeb archaeolegol diwydiannol, a rhyw gymaint o ddefnydd adloniadol. Yr ymarfer amgylcheddol gorau fyddai canolbwyntio defnyddio/ail-ddefnyddio gwastraff llechi ar safleoedd gweithfeydd presennol ac ar domenni gwastraff lle na fyddai fawr o broblemau amgylcheddol, neu lle byddai manteision clirio'r gwastraff llechi yn gorbwyso'r problemau. Gallai clirio dewisol, ar raddfa fechan ac ar gyfer defnydd lleol olygu llai o angen cludo defnyddiau o rywle arall, a gallai hyn fod yn rhan o gynllun adfer i baratoi'r tir at ddefnydd arall.

Mae chwareli sydd yn gweithio ym Mlaenau Ffestiniog a Bethesda, Gwynedd ac yn Kirkby in Furness, Cumbria yn agos at reilffyrdd neu borthladdoedd. Gydag ychydig o gymorth ariannol, gellid cynhyrchu sawl miliwn o dunelli'r flwyddyn a'u cludo i gyfrannu at y galw cenedlaethol am gerrig mâl.

Defnydd newydd ar hen weithfeydd llechi

Gwneir defnydd newydd o lawer o hen weithfeydd llechi bellach gan y cyhoedd, mewn grwpiau neu fel unigolion, yn aml iawn heb unrhyw baratoadau ar y safle, na hyd yn oed gael caniatad y perchennog.

Mae rhain yn cynnwys:
- cerdded ac adloniant anffurfiol;
- astudio archaeoleg diwydiannol, bywyd gwyllt, daeareg;
- gweithgareddau awyr agored, megis dringo, abseilio a nofio tanddwr.

Mae teithwyr wedi meddiannu ambell safle. Mae eraill yn cynnwys hen geir a adawyd a gwastraff a dipiwyd ar y slei.

Gwelwyd defnydd anffurfiol a goddefol, megis coetir a thir pori yn ystod yr ymchwil. Mae defnydd ffurfiol yn debygol o fod angen gwaith i baratoi'r safle ac i glirio peryglon i ddefnyddwyr y safle. Bydd gwaith sydd yn golygu datblygiad neu weithgarwch peirianyddol angen caniatad cynllunio. Mae safleoedd wedi cael eu datblygu fel stadau diwydiannol a stadau tai, canolfannau twristiaeth a meysydd parcio. Rhoddwyd caniatad i gael gwared ar wastraff diynni ar safleoedd, ond go brin y caniateir gwaredu gwastraff domestig gan ei bod yn anodd cwrdd â'r gofynion i reoli a thrin hylifau fyddai'n trwytholchi drwy'r gwastraff.

Dulliau adfer

Yn y cyfnod rhwng 1972 a 1992, cwblhawyd un ar bymtheg o gynlluniau adfer tir as safleoedd diffaith, gyda chymorth grantiau adfer tir. Roedd pob un o'r cynlluniau hyn ar y cyfan neu yn gyfangwbl yn cwrdd â'r amcanion a osodwyd iddynt, er y byddai mwy o waith yn gwneud iddynt gymhathu'n well yn y tirwedd o'u cwmpas.
Y prif elfennau oedd:

- clirio adeiladau a rwbel;
- gwyro neu glirio dwr;
- cloddio ac ail leoli gwastraff lechi;
- cywasgu'r deunydd llanw;
- trimio'r wynebau a ailraddiwyd;
- gwasgu, paratoi a hadu;
- darparu draenio dwr wyneb;
- plannu, ffensio a chynnal a chadw'r tirwedd;
- daparu'r isadeilaeth ar gyfer defnydd newydd.

Yn y rhan fwyaf o'r cynlluniau, cliriwyd cyfansymiau mawr o wastraff llechi gan ddefnyddio cyfuniad o dechnegau peirianneg sifil a chwarelydda i ddelio â'r deunydd mwyaf bras. Gosodwyd y llechi mewn haenau a'i gwasgaru â pheiriannau lefelu. Lle trefnwyd datblygu, cywasgwyd y wyneb â roleri. Mae'r dulliau hyn wedi rhwystro'r deunydd mewnlenwi rhag ymsefydlogi. Ymhob un o'r cynlluniau, gwnaed darpariaeth i gasglu a chael gwared ar ddwr wyneb a'r dwr draenio a all ddeillio o waelod tomenni gwastraff llechi yn ystod glawogydd trwm cyson. O ddiffyg pridd wyneb, paratowyd lleiniau mawr ar gyfer hau glaswellt trwy wasgu'r wyneb â roleri grid trwm. Roedd hyn yn rhoi haen denau o ronynnau llechi manach y gallai glaswellt gydio ynddo pe rhoddi nitrogen a ffosfad ynddo. Roedd y tir glas hwn yn agored i gael ei niweidio trwy bori direolaeth ac am nad oedd yn cael ei wrteithio'n rheolaidd. Mae dirywiad y tir glas yma yn dangos angen cydbwyso'r amcanion defnydd tir a'r dewis o rywogaethau â'r adnoddau sydd ar gael i'w rheoli. Sefydlwyd coed yn llwyddiannus mewn gwelyau o bridd, mewn haenau o lechi a falwyd yn fân, ac mewn pocedi o gompost. Gwelwyd fod rhywogaethau sydd yn gwladychu'n naturiol mewn tommeni gwastraff, megis Bedwen, Cerddinen, Helygen Grynddail wedi gwneud yn dda, am eu bod yn medru gwrthsefyll sychder, diffyg maeth a bod yn agored i'r tywydd. Os gellir dal ati i fwydo nitrogen i goed, llwyni a legiwmiau mewn coetiroedd athiroed glas, gellid cael gwell tyfiant ymhob rhywogaeth. Ni all prosiect adfer lwyddo oni bai fod y llystyfiant a sefydlir yn medru ffynnu dan yr amodau defnydd a chynnal a chadw fydd yn dilyn y cynllun cychwynnol.

Asesu deiwsiadau defnydd tir ac adfer

Mae i safleoedd gweithfeydd llechi nifer o ffactorau, negyddol a chadarnhaol, y dylid eu hystyried wrth ddewis defnydd newydd ar gyfer safleoedd ac wrth baratoi cynlluniau diweddaru ar gyfer adfer dan ganiatadau cynllunio presennol. Gall safleoedd gwrdd â rhywfaint o'r anghenion defnydd tir mewn ardal os gwneir mân addasiadau priodol. Gosodir fframwaith ar gyfer asesu safle sydd yn anelu at gynorthwyo'r defnyddiwr i gymharu safleoedd posib â'r gofynion defnydd tir y gelwir amdanynt. Mae'r fframwaith yn cynnwys asesiad ar ba mor ymarferol fyddai clirio gwastraff llechi, fel y gellid ystyried defnydd tir ar y safle, heb ei aflonyddu neu ar ei newydd wedd.

Mae asesu safle yn broses aml-disgyblaeth sydd yn cynnwys:-

- sefydlogrwydd y tomenni, y chwareli a'r adeiladau;
- potensial ar gyfer defnyddio'r gwastraff llechi;
- pa mor hygrych yw'r safle;
- draeniad;
- gwerth y safle o ran bywyd gwyllt;
- asesu'r tirwedd;
- diddordeb daearegol;
- potensial adloniadol;
- gwerth y safe o ran archaeoleg diwydiannol;
- potensial adfer y safle.

Datblygwyd ymagwedd i asesu tirweddau'r safleoedd gweithio llechi'n sustematig fel rhan o'r ymchwil a chyflwynir ef yn yr adroddiad hwn.

Astudiaethau áchos

Mae'r adroddiad yn cynnwys saith astudiaeth achos, a dynnwyd o Ogledd Cymru, Ardal y Llynnoedd a'r Alban. Mae'r astudiaethau achos hyn yn dangos materion megis:-

- rhelaeth cynllunio ac adfer safleoedd;
- gwrthdaro â bywyd gwyllt a thirweddau gwerthfawr;
- ffynonellau mwynau eilradd, a chanlyniadau amgylcheddol cloddio amdanynt;
- twristiaeth, archaeoleg diwydiannol a chadwraeth;
- gwaredu gwastraff;
- adfer a rheolaeth tymor hir;
- defnyddiau anffurfiol, yn cynnwys coetiroedd, cerdded, dringo ac abseilio.

1. The slate workings in Britain range from excavations just a few metres deep, to operations such as Delabole Quarry, north Cornwall seen here. The hole is over 150m deep and 1.6km in circumference.

PHOTOGRAPH COURTESY OF RTZ MINING & EXPLORATION LTD.

2. Modern techniques of slate extraction at Penrhyn Quarry, Bethesda may be seen alongside centuries-old working faces.

PHOTOGRAPH COURTESY OF ALFRED MCALPINE SLATE PRODUCTS LTD.

3. Deterioration in slate retaining walls requires careful monitoring and, often, remedial action. Failure of the wall would release much of the tip, burying an access road.

4. Massive abutments remain at many workings. This one, in danger of collapse, was removed during the reclamation scheme at Talysarn.

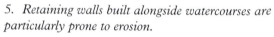

5. Retaining walls built alongside watercourses are particularly prone to erosion.

6. *Details such as walls should be designed to respect local styles: north Cornwall and north Wales.*

7. *Extraction from the base of a waste tip can leave an unstable face which may collapse if disturbed.*

8 *Flat topped slate waste tips provide the only working space at many quarries in mountainous terrain.*

PHOTOGRAPH COURTESY OF THE LAKE DISTRICT NATIONAL PARK AUTHORITY.

9. *Suitable slate waste may be crushed and screened to produce a range of construction aggregates, granules and powders.*

10. *Compaction of the slate waste used as fill material. Blaenau Ffestiniog 1978.*

11. *The new A487(T) alignment was prepared by regrading and compacting the slate waste tip. Corris 1978.*

12. Bowithick Quarry, north Cornwall, was filled with domestic refuse and capped.

13. Dorothea Quarry is a popular venue for divers, although its great depth leads to serious decompression accidents. See case study 3.

14. A simple reclamation scheme at these lakeside slate waste tips improved access and safety for recreational users. Y Glyn, Lanberis.

15. *Trees planted in beds of subsoil have grown rapidly, while birch, gorse, heather and other species have colonised the slate waste surface. Corris 1994. See photograph 11.*

16. *Experimental work carried out to develop simple revegetation techniques. Fencing, to exclude sheep and rabbits, is essential.*

17. *This embankment was constructed using slate waste. Pocket planting was successfully used to establish the trees, without any treatment of the slate waste surface. Abercwmeiddau, 1993.*

18. *Recolonisation and remoteness from human interference, make sites such as this attractive for wildlife. Aberfoyle, Stirling.*

19. *Many quarries are not fenced. Walkers in mountainous areas may be prepared for such hazards, but visitors to lower open spaces are less likely to be aware of the dangers. Moel y Faen Quarry, near Llangollen.*

20. *The majority of visitors to this popular site do not regard the quarries or waste tips as an eyesore. Horseshoe Pass, near Llangollen.*

1 Introduction

1.1 Background

1.1.1 Slate workings in Britain

Slate has been worked in Britain for many centuries, providing roofing and building materials. The area of land affected by slate extraction is small compared with many other mineral types, but vast quantities of waste materials have been generated. It is estimated that there are approximately 500 million tonnes of slate waste in Britain and that 6 million tonnes are added each year. Virtually all of this waste is disposed of in waste tips.

Many slate workings are within or visible from National Parks or Areas of Outstanding Natural Beauty, and so the visual impact of these workings can be significant despite the relatively small area involved. The quarries, often deep steep-sided excavations, raise concerns over public safety, but also represent the cultural and industrial heritage of communities.

1.1.2 Definitions

The term 'slate' is widely applied to rocks which can be split into relatively thin sheets or slabs, suitable for roofing and other purposes. The geological definition of slate, adopted for this project, is a fine-grained sedimentary rock such as mudstone, siltstone, volcanic ash or tuff which has been metamorphosed (altered) by heat and/or pressure and has recrystallised in the solid state. The resulting alignment of minerals within the rock mass gives slate its property of fissile cleavage and hence its practical and economic value.

The term 'slate' has historically been applied to other forms of sedimentary rock which can be split to form useful products including roofing tiles, but these minerals rely on cleavage along the original bedding plane rather than on mineral metamorphism. Consequently they cannot be split as thinly as 'true' slates and are relatively heavy as a roof covering. Examples of true slates, and other minerals known as stone slates or flagstones, are given in Table 1.1. These other minerals were used in the localities where they could be worked but were not widely transported to urban areas. The industry remained small, although some notable buildings have roofs covered with stone from these areas.

This research project has concentrated on the 'true' slate workings of Cumbria and Cornwall in England; Scotland, and Wales. There are also some small abandoned workings of Cambrian slates on the Isle of Man, and of pre-Cambrian slates around Swithland, north-west of Leicester. However these workings have not specifically been included in the study, but many of the findings in this report concerning smaller slate workings may be applicable to these other workings.

In this report the term 'workings' is used for all slate extraction sites, whether opencast or underground, and the associated waste tips.

For the purposes of this report:

'Restoration' is taken to mean works carried out by a quarry operator to restore tipped, quarried or disturbed land to some new use or vegetation cover. This work may be carried out voluntarily; in response to planning permission conditions; or to provide new

Table 1.1 *Geological summary of some slates, flagstones and roofing materials*

Location	Geological formation and rock type	Location	Geological formation and rock type
Delabole, Cornwall	Devonian Slate	Caithness (flagstones)	Devonian Old Red Sandstone
Kirkby, Cumbria	Silurian Slate	Derbyshire (roofing)	Carboniferous Sandstone and Millstone Grit
Ffestiniog, Gwynedd	Ordovician Slate		
Penrhyn, Gwynedd	Cambrian Slate	Collyweston, Leicestershire (roofing)	Jurassic Oolitic limestone
Ballachulish, Highland	pre-Cambrian Slate	Cotswold (roofing)	Jurassic Oolitic limestone

working space for the operation. It includes "aftercare" as defined in the Town and Country Planning Act 1990.

'Reclamation' is taken to mean works carried out by a second party to treat abandoned workings for which there is no prospect of restoration by the former operator.

1.1.3   Options for the future

Substantial schemes have been carried out with central Government grant-aid to reclaim some derelict slate workings, but such radical treatment is not appropriate for all slate workings. Increased appreciation of the value of these sites for wildlife, industrial heritage and other characteristics has generated new interest in sensitive schemes to provide opportunities for education and recreation as part of site rehabilitation.

There is potential and government encouragement for the increased use of slate wastes as a secondary aggregate and fill material for construction. The large-scale removal of slate wastes could aid reclamation in some cases but would cause environmental problems in others which might outweigh the benefits of reclamation.

Very few slate workings and tips currently have any provisions under planning control for their eventual restoration. There are occasional opportunities to introduce such provisions as old planning permissions are replaced by new. The Department of the Environment consultation paper 'The Reform of Old Mineral Planning Permissions' (DOE 1994a) sets out proposals which would require operators to put forward new schemes of working and restoration for active or re-activated workings having planning permissions granted between 1948 and 1981. If these proposals become law, many more workings will eventually be restored by the slate industry. However, the future treatment of abandoned slate workings will probably lie outside the active slate industry.

1.2   Research for the Department of the Environment

1.2.1   Objectives of the research

Against this background the Department of the Environment commissioned research to examine the strategic and technical options for the future management of current arisings of slate waste and of old tips and quarries. The research, which forms the basis of this report, set out to assess the need and potential for reclaiming slate waste tips and quarries in Britain, and to provide technical guidance on the methods available. It sought to identify and assess land use planning opportunities and constraints for increasing the use of slate waste in applications such as aggregate and fill material, cement and brick-making, and to provide guidance to central and local

government and the slate industry to assist in the selection of priorities for reclamation, restoration and new site uses. This research complements projects to study the technical aspects of slate waste as a construction material (BRE 1994) and the occurrence and utilisation of various mineral wastes (Arup 1991).

1.2.2   Research methods

In order to meet the objectives described, the research team gathered and reviewed information by consulting statutory and other interested organisations, searching published and unpublished material, visiting sites throughout Britain, and meeting representatives of local government, the industry and many other bodies with interests in slate working sites.

A Pilot Study was conducted to develop and test a method for the assessment of land use and reclamation options for areas of slate workings. The method was tested by applying it to the slate workings around Blaenau Ffestiniog, Gwynedd.

This final report was prepared to present the findings of the research to a wider audience. The report contains, in section 2, a description of slate workings and the issues addressed in the research and, in section 3, a discussion of the consequences of greater use of slate waste. Some alternative uses of former slate workings are suggested and described in section 4, and in section 5 approaches to reclamation are discussed together with the techniques which are available. A refinement of the method developed for the assessment of land use and reclamation options is presented in section 6. The issues are exemplified in the Case Studies in Annexe 1, and illustrated by further photographs in Annexe 2. A list of organisations and individuals consulted during the research forms Appendix 1 to this report; the sites visited are listed in Appendix 2, and all references to documentary or personal sources of information are detailed in full in the reference list.

Although available drafts of some forthcoming documents have been reviewed, policy and technical guidance continues to develop and so elements of this report may in time be superseded.

2 Slate Workings in Great Britain

2.1 The scale of the industry

2.1.1 Production and markets

The annual production of slate in Britain in the period 1985-1991 averaged some 41,000 tonnes of roofing, cladding and decorative materials, and 43,000 tonnes of powders and granules. A further 283,000 tonnes per annum (tpa) on average, of slate waste was sold for fill and other aggregates uses in the period (DOE 1992a).

Natural slate holds around 1% of the UK roofing market in the housing sector, and 4% in the commercial sector (NSQA pc). Domestic roofing slate consumption grew from 37,000t in 1985 to 85,000t in 1990. Imported slate supplied almost two-thirds of this consumption (Harries-Rees 1991). Although statistics are not yet available, it is known that the recession in the building industry has led to substantially reduced demand in 1991 and 1992, and

industry forecasts are that some 55,000 to 60,000 tpa will be required over the next 15 years. Approximately half of this will be supplied by UK producers and half by imported slate. Table 2.1 shows the forecasts for other sectors of the slate market. The market for roofing slate is considered to be mature, that is, that the proportion of roofing carried out using slate will remain constant and the size of the market for slate will grow or contract in proportion to the overall building market (Hughes B. pc).

2.1.2 Employment

At its peak in 1898, the North Wales slate industry produced over 500,000 tonnes of slate and employed over 16,000 men. The industry declined rapidly until the 1970s, but expanded in the 1980s following increased demand and substantial investment in modern production methods. The long decline in slate production is illustrated in Table 2.2.

Table 2.1 *Forecasts of the demand for slate products, by market sector 1995-2000*

Market sector	Forecast UK demand, tonnes per annum	Forecast market share for UK producers	Forecast UK exports, tonnes per annum
Roofing slate	56 700	53%	2 500
Architectural slate	6 200	70%	1 000
Slate walling	1 860	100%	-
Slate paving	3 100	100%	500
Slate powders and granules	72 170	98%	500
Aggregates	208 000	100%	-
Bulk fill	103 000	100%	-

DATA SUPPLIED BY THE NATURAL SLATE QUARRIES ASSOCIATION

Notes:
Annual demand and export data are forecast averages over the five year period.
UK demand in all sectors is expected to grow in proportion to the building market.
Market share for UK producers is expected to remain constant.
UK exports of roofing slate are expected to grow by about 3.5% per annum over the next 10-15 years.

Table 2.2 *Employment and output in the slate industry*

A. The slate industry in 1882

Region	No of concerns	Employed	Output (tons)
Argyll	4	540	17 990
Perth	1	3	75
Scotland	**5**	**543**	**18 065**
Lancashire	6	86	2 574
Westmorland	8	107	1 580
Cornwall	6	410	11 680
Devon	3	54	1 378
Yorkshire	7	89	6 744
Durham	2	12	390
England	**32**	**749**	**24 346**
Caernarfonshire	36	8 960	280 716
Merionethshire	29	5 086	166 601
Denbighshire	3	162	8 515
Montgomeryshire	2	51	940
Wales	**70**	**14 259**	**451 689**
Britain	**107**	**15 551**	**494 100**

DATA FROM 'THE WELSH SLATE INDUSTRY', MINISTRY OF WORKS (1947)

B. The slate industry in 1992

Region	Employment 1992_1	Employment peak 1988-1992	Output (tonnes) 1992_2
Cumbria	190	260	3 650
North and Mid Wales	530	630	26 250
Cornwall and Devon	70	90	400
Britain	**790**	**980**	**30 300**

Notes:
1 Data from discussions with operators in late 1992/early 1993. Rounded to the nearest 10. Numbers are those directly employed, related employment eg in haulage is not included.
2 Approximate output of roofing slates in 1992.

At the end of 1992 the slate industry employed approximately 790 people nationally in extraction, production, sales and support activity (quarry managers, pc). Although this number is small in relation to the UK aggregates or minerals industry as a whole, the employment is highly significant locally since the industry is concentrated in small areas with relatively few other major employers.

Production is now concentrated in three areas of Britain:
- southern Cumbria;
- north Cornwall;
- north and mid Wales.

2.1.3   Cumbria

In Cumbria, Burlington Slate Ltd operates Kirkby Quarry at Kirkby in Furness, and six 'satellite' quarries within the Lake District National Park. All extracted material is transported to Kirkby Quarry for processing. Kirkstone Quarries Ltd operate Petts Quarry, and small scale extraction continues at Horse Crag, Hodge Close and Brathey Quarries. Cumbria supplies about 12.5% of the UK output of roofing slate. Other dormant sites such as Honister Crag Quarry could recommence small scale working if the market warranted it.

Cumbrian slates range in colour from silver grey, and light to dark green Ordovician slates of volcanic ash origin known as 'green slate', to blue-black and grey Silurian slates of clay minerals (Anon, 1986). Lakeland buildings are traditionally roofed in the 'green' slate, and continued demand exists for slates in the range of colours produced locally. Demand is reinforced by development control policies of the National Park Authority which ensure that the character of local buildings is maintained. Green slate is also manufactured into a range of high-value decorative and architectural products for domestic and export markets.

2.1.4   North Cornwall

In north Cornwall the major producer is Delabole Slate Ltd, owned by RTZ Plc. Devonian grey slate is extracted and processed into roofing slates, dimension stone, walling and paving stone, and granules and powders. The output of roofing slate is approximately 1.5% of the UK total. Smaller scale quarries are operated in the coastal strip, producing iron-stained 'rustic' slate for paving and walling stone. Similar quarries operate near Tavistock, Devon.

2.1.5   North and mid Wales

In north Wales, production continues in the Bethesda-Nantlle Cambrian slate belt which is dominated by Penrhyn Quarry at Bethesda. Penrhyn Quarry is operated by Alfred McAlpine Slate Products Ltd, and supplies more than half the total UK roofing slate production. Smaller scale production continues at Pen yr Orsedd in the

Nantlle Valley. Cambrian slates vary in colour from green to blue-grey and purple.

The slates of the Ffestiniog belt are uniformly blue-grey in colour, and are of Ordovician age. They are quarried by the Ffestiniog Slate Quarries group of companies, and by Greaves Welsh Slate Company at various sites in the vicinity of Blaenau Ffestiniog, to produce roofing slates and architectural and dimension stone products. Collectively, the production sites in this area account for 35-40% of the UK output of roofing slate.

Slates of the Ordovician series are also extracted from one underground mine at Aberllefenni near Corris, in mid Wales. As the material does not split well, the company, Wincilate Ltd, specialises in the production of sawn slabs for architectural, monumental and decorative work.

2.1.6   Slate in conservation work

Slate is a highly variable material, and production methods evolved in response to this. As a result local styles of slate roofing developed and many colours, shapes and sizes of slate have been used on historic buildings. Active slate quarries play an important part in the maintenance of local building character in, for example, the Lake District and Snowdonia National Parks and in older areas of many cities where continued supplies of precise types of slate are essential for renovation and new roofing work. The use of slate of specific appearance is a strict planning policy within these National Parks.

2.1.7   Slate waste

The production of roofing slates generates large amounts of material of poor quality which cannot be split and sold. There is a great variation in the ratio of waste to useful slate but on average over 20t of waste is generated for each tonne of slate that is sold.

All producers have sought to increase their efficiency and to sell a greater proportion of the slate they extract from the quarries. Historically, only roofing slate was worth processing for sale and so vast waste tips were deposited. Today, producers aim to convert as much of the raw slate block into saleable slate products as they can, and so material is graded and sorted at all stages. Most producers market a range of products, including sawn block derived from material which will not split well, and paving or walling stone from rejects. The larger producers, Delabole Slate Ltd ('Delabole') and Alfred McAlpine Slate Products ('McAlpine'), together with Redland Aggregates Ltd ('Redland') also produce powdered slate and slate granules for specialised applications such as fillers for bitumastic products, and reconstituted slate tiles. Delabole are able to sell virtually all the material extracted from the quarry by this means, while McAlpine have an integrated production pattern which utilises waste from one product line as the raw material for another, giving considerable economies of production. Redland sited their plant at an active slate quarry to utilise its slate waste, but also import slate wastes to provide a range of slate colours. These 'other uses' of slate and slate waste are discussed in detail in section 3 Slate waste as a resource.

Most producers also sell slate waste for uses such as bulk fill, but only in response to specific enquiries. There is generally insufficient local demand for regular sales. McAlpine produce specific construction aggregates and sell an average of 250,000 tonnes/yr to local markets. This should be seen in the context of their annual waste production of approximately 3,000,000 tonnes (Hughes B. pc).

2.2 Extent and location of abandoned workings

2.2.1 Definitions

For the purposes of this report, 'abandoned' is taken to mean those workings outside the operational areas of currently active sites, at which no further extraction/processing/tipping or other use is proposed. Some of these workings could be the subject of renewed interest for slate extraction in the future, should market conditions change. 'Dormant' is used to mean workings which retain valid

planning permission for working, but are not currently producing slate. 'Derelict' is used in the sense of the administrative definition adopted for derelict land grant purposes by the Department of the Environment (1991a) and Welsh Development Agency, ie "land so damaged by industrial or other development that it is incapable of beneficial use without treatment". This definition includes worked out mineral excavations and tips, but excludes land which is subject to enforceable planning conditions or other arrangements providing for restoration. Land still in use such as an active tip whether or not there are provisions for restoration, and land which has blended into the landscape or has been put to some acceptable use so that it no longer constitutes a problem is also excluded.

2.2.2 The extent of derelict workings

The data collected for the main slate working areas in the three national surveys is summarised in Table 2.3.

As part of the 1988 Derelict Land Surveys, local authorities were asked to identify land which although derelict did not justify reclamation due to remoteness, lack of suitable afteruse or impracticality of treatment.

Figure 2.1 shows the location of this derelict land. The county planning authorities collated the survey data and

Table 2.3 *Derelict slate workings in Britain*

Area	Derelict slate workings justifying some treatment hectares	Total dereliction hectares
Scotland		
Lochaber LEC	N/A	36
Argyll and Islands LEC	N/A	4
Forth Valley LEC	N/A	371
Dunbartonshire LEC	N/A	324
England		
South Lakeland District, Cumbria	12	36
Wales		
Preseli District, Dyfed	17	70
Arfon Borough, Gwynedd	568	624
Dwyfor District, Gwynedd	1	72

Sources:
1. Scottish Vacant and Derelict Land Survey 1990 (Scottish Office, 1992). Data recorded on Local Enterprise Company (LEC) basis. Slate not identified separately.

2. Survey of Derelict Land in England 1988 (Department of the Environment, 1991). Excludes many workings not justifying treatment and not recorded.

3. Survey of Derelict Land in Wales (unpublished data from WDA). Excludes many inactive workings with valid planning permission, and 17 ha in current WDA reclamation programmes.

Figure 2.1 Active and derelict slate workings, 1994

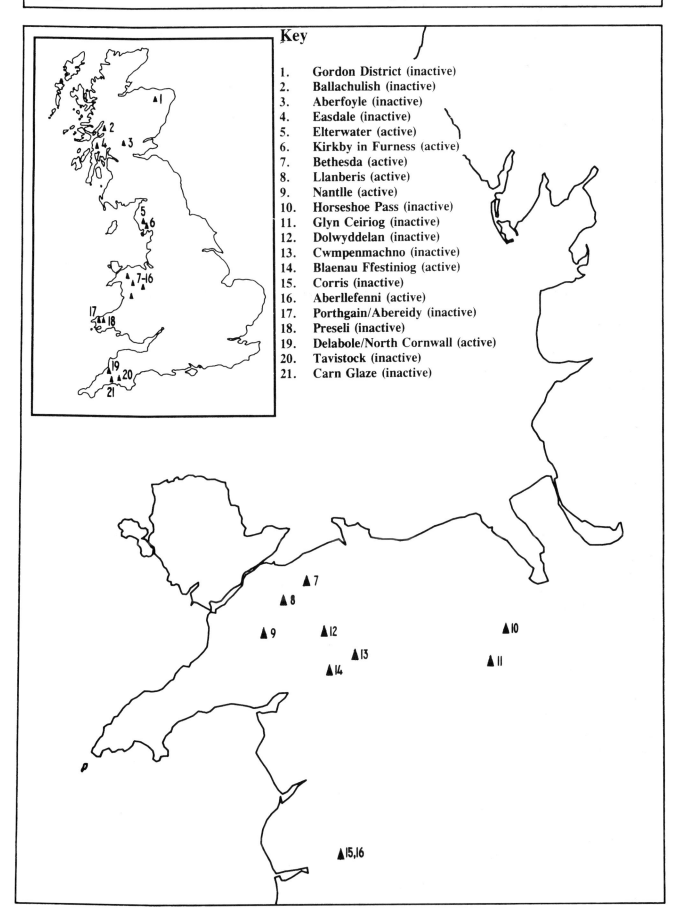

Key

1. Gordon District (inactive)
2. Ballachulish (inactive)
3. Aberfoyle (inactive)
4. Easdale (inactive)
5. Elterwater (active)
6. Kirkby in Furness (active)
7. Bethesda (active)
8. Llanberis (active)
9. Nantlle (active)
10. Horseshoe Pass (inactive)
11. Glyn Ceiriog (inactive)
12. Dolwyddelan (inactive)
13. Cwmpenmachno (inactive)
14. Blaenau Ffestiniog (active)
15. Corris (inactive)
16. Aberllefenni (active)
17. Porthgain/Abereidy (inactive)
18. Preseli (inactive)
19. Delabole/North Cornwall (active)
20. Tavistock (inactive)
21. Carn Glaze (inactive)

hold 1:50,000 or 1:10,000 scale maps showing individual sites.

There is a further 600ha of land left derelict by slate working (Table 2.3) and 1500ha of land which without treatment is likely to become derelict when working ceases (Table 2.4).

Land reclamation schemes for abandoned slate workings and tips have been carried out at:
- Ballachulish, Highland Region, west Scotland (25ha);
- Thirteen sites in north and mid Wales (270ha);
- Jeffries pit and Prince of Wales Quarry, north Cornwall (1ha).

The approaches adopted in these schemes are described in section 5.

2.3 The extent of slate workings with planning permission

2.3.1 Active slate workings

Over one thousand hectares of land in Britain have been disturbed by slate working, and planning permissions exist for a further five hundred hectares (see Table 2.4). Many sites have been in operation since the eighteenth or nineteenth centuries. According to surveys carried out by mineral planning authorities for the DOE and the Welsh Office, very little of this land, just 31ha, is covered by satisfactory conditions requiring the restoration of the land. The remainder of the land is worked under Interim Development Order permissions dating from 1943-48 or early planning permissions with few, if any, conditions governing the eventual restoration of the site.

Very few planning permissions for slate extraction have been granted in recent years, and consequently there have been few opportunities to introduce modern enforceable conditions requiring comprehensive site restoration. Opportunities have arisen through the review of IDO permissions, covering two large slate operations, and through the submission of 'consolidating' applications in which existing workings have been taken together with proposed extensions of extraction or tipping. Such opportunities may continue to arise, in some cases because operators are seeking disposal sites near the sites of extraction in order to reduce the cost incurred in transporting slate waste to more remote tipping sites.

Restoration work by the industry has, to date, been extremely limited in extent. The reasons for this include:
- lack of any requirements for restoration;
- lack of planning control until modern times;
- limited ability of mineral planning authorities to update permissions still operating;
- continuing GDO tipping permissions covering many old tips;
- assumption until recently that restoration was impractical or unacceptably expensive;
- belief that slate waste tips would ultimately become a resource;
- concern that future extraction or tipping areas would be 'sterilised' by restoration;
- lack of topsoil in new extraction areas for use in restoration.

The extent of operator-led restoration found during the study is as follows:

1. Petts Quarry, Cumbria. Kirkstone Quarries Ltd, with professional advice, began a programme of tipping subsoil, manure, roadside sweepings (principally leaves), fine-grained waste slate and similar materials over the surface of completed tipping areas. Grass seeding was carried out in the autumn of 1992, but growth was expected to be slow due to the exposure and altitude (520-550m AOD). Future tipping will be in 'lifts' which will in turn be grassed. Ultimately the tip may be returned to mountain grazing (Fecitt N. pc).

2. Gloddfa Ganol Slate Mine, Blaenau Ffestiniog. The owners have restored a 5-10 metre wide strip along part of the visitor access road to the mine. Soil and peat from within the site were spread for grass seeding, tree and shrub planting. The exercise is partly intended to show what will be done elsewhere at the site when opportunities arise.

3. Elsewhere a number of operators have carried out 'ad-hoc' works such as tipping fine-grained slate waste, subsoil or overburden onto tip faces to encourage natural regeneration, but generally without expert guidance or any overall plan.

Recent planning permission conditions have required either specific works or that a restoration plan is submitted. During the study, no evidence was found that such requirements had yet been implemented.

2.3.2 Dormant workings

Slate extraction rarely, if ever, proceeds systematically through a deposit until the reserve is exhausted. Variation in the mineral quality and in market demand leads to intermittent working of some quarries. Parts of sites or complete sites may lie dormant for years or even decades, but be reactivated when the demand for that particular mineral exists. Slate workings are, therefore, rarely backfilled and restored.

The introduction of modern working methods can allow profitable slate production from sites which have been inactive for decades. Where valid planning permissions remain, working could be resumed at any time until the terminal date of the permission. Such land is not therefore included in the derelict land statistics, and the possibility of future

Table 2.4 *Slate workings in Britain, April 1988*

	Land covered by permissions with no reclamation conditions nor any other arrangements for reclamation				Land covered by permissions which include conditions or arrangements providing for reclamation			
	Active			Ceased	Satisfactory conditions or arrangements for reclamation			
	Total area worked	Total area not worked	Total of 1 & 2	Total area worked	Total area worked	Total area not worked	Total of 5 & 6	Permission for underground mining
	1	*2*	*3*	*4*	*5*	*6*	*7*	*8*
Extraction								
Cumbria including Lake District	36	171	207	0	10	3	13	0
Cornwall	7	45	52	2	5 (6)	1 (3)	6 (9)	0
Devon	0	0	0	0 (12)	1 (5)	0 (17)	1	0
Surface tipping								
Cumbria including Lake District	15	32	47	3	6	4	10	0
Cornwall	6	0	6	0	0	0	0	0
Devon	0	0	0	3	1	0	1	0
England sub-total	64	248	312	8	41	16	57	0
Extraction								
Clwyd	2	7	9	0	0	0	0	0
Dyfed	6	3	9	0	13	1	14	0
Gwynedd	224	41	265	15	2	0	2	374
Surface tipping								
Clwyd	7	18	25	0	0	0	0	0
Dyfed	0	0	0	0	0	0	0	0
Gwynedd	499	177	676	109	4	0	4	0
Wales sub-total	738	246	984	124	19	1	20	374
Britain total	802	494	1,296	132	60	17	77	374

All data in ha. No regular working takes place in Scotland.
Column 4 includes 3ha in the Lake District where restoration conditions are likely to remain unfulfilled.
Columns 5-7 include in brackets land for which restoration conditions are unsatisfactory.

DATA FROM SURVEY OF MINERAL WORKINGS IN ENGLAND 1988   HMSO
      RESULTS OF THE SURVEY OF LAND FOR MINERAL WORKINGS IN WALES   WELSH OFFICE 1988

mineral working may deter landowners from seeking alternative uses or restoration of these sites.

Slate waste tips are often considered by their owners to be an asset for future sale as fill material or for another beneficial use. This consideration may become an obstacle to reclamation or restoration to provide new uses for the land.

2.4 Planning controls over mineral workings

2.4.1 Introduction

The comprehensive land use planning control over mineral working dates from 1 July 1948 when the Town and Country Planning Act 1947 introduced general planning control over the development of land. There have subsequently been revisions of this Act, the latest consolidation producing the Town and Country Planning Act 1990, but the basic framework of forward planning and of development control has remained substantially the same. The special features of mineral development have been recognised in several features of the main Town and Country Planning legislation and in secondary legislation such as the General Development Order and in Regulations. Government guidance and information on this legislation is set out in the series of Planning Policy Guidance Notes (PPGs) and particularly Minerals Planning Guidance Notes (MPGs) published by the Department of the Environment and the Welsh Office. The Scottish Office have also published guidance in National Planning Policy Guidance 4 (SO 1994).

2.4.2 The aims and means of control

The Minerals Planning Guidance Note 1 (MPG1) 'General Considerations and the Development Plan System' (DOE 1988a) sets out the aims of planning control as it applies to minerals as:
- to ensure that the needs of society for minerals are satisfied with due regard to the protection of the environment;
- to ensure that any environmental damage or loss of amenity caused by mineral operations and ancillary activities is kept to an acceptable level;
- to ensure that land taken for mineral operations is reclaimed at the earliest opportunity and is capable of an acceptable use after working has come to an end;
- to prevent the unnecessary sterilisation of mineral resources.

The key elements of the means of control are given as:
- structure plans which set out policies and general proposals for minerals within the national context and provide for the coordination of mineral working with other elements of strategic planning;
- local plans which develop the policies and general proposals of structure plans and relate them to identifiable areas of land;
- unitary development plans, which cover London and the former metropolitan county council areas (none of which contain slate workings);
- the grant or refusal of planning permission for the working of minerals in any particular land, for the erection of associated plant or buildings, for the disposal of mineral waste or for other ancillary purposes and the imposition, when planning permission is granted, of conditions;
- the enforcement of planning control to prevent unauthorised development and to ensure compliance with planning permissions.

In addition the local authorities have statutory powers enabling them to enter into agreements for regulating the use or development of land *eg* section 106 Agreements.

2.4.3 Areas given special attention

Many slate workings in Britain are in or near National Parks, areas of high landscape value or other environmentally significant areas, although such designations postdate the original establishment of most of these workings. According to a Government statement in both Houses of Parliament in April 1987 all proposals for mineral working in these areas should "be subject to the most rigorous examinations" and all local authorities should have regard to this policy in the preparation of, and alterations to, their development plans.

2.4.4 Minerals content of development plans

The Town and Country Planning Act 1990 requires local planning authorities to prepare Structure Plans which shall "..include policies in respect of-
  (a) the conservation of the natural beauty and amenity of the land;
  (b) the improvement of the physical environment; and
  (c) the management of traffic."
"In formulating their general policies the authority shall have regard to-
  (a) any regional or strategic planning guidance given by the Secretary of State to assist them in the preparation of the plan;
  (b) current national policies;
  (c) the resources likely to be available; and
  (d) such other matters as the Secretary of State may prescribe ..".

Planning Policy Guidance Note 12 and Mineral Planning Guidance 1 advise that development plans should take account of the following:
- national and regional policies;
- safeguarding of deposits bearing in mind that a mineral can only be worked where it exists;

- securing the extraction of a mineral to supply local, regional and national requirements whilst minimising harm to the environment;
- proposals to ensure the reclamation and beneficial afteruse of old mineral workings;
- providing for sufficient stock of permitted reserves;
- identifying areas for future working bearing in mind other land uses;
- minimisation of disturbance from transportation of minerals, the consideration of methods and routeing;
- ancillary operations including waste disposal;
- production of a check list of considerations relevant in assessing applications for working;
- considerations relating to agricultural land quality, National Parks, Areas of Outstanding Natural Beauty, Nature Reserves and Sites of Special Scientific Interest.

MPG 1 (DOE 1988a) advises with regard to the national policy for slate working, that, "It is important to recognise that in some cases it is quarried from geological formations which are very restricted in occurrence. There is often a large proportion of waste and production can be intermittent, plant and employee cost per tonne of production is usually higher than for other quarries. It should also be borne in mind that there may be cases where the working of dimension stone and slate reserves can be expected to continue, sometimes intermittently, for very long periods indeed. Also that working and processing generally involve smaller acreage and lower production rates than any other mineral operations".

Development plans took on greater significance in the consideration of planning applications with the passing of the Planning and Compensation Act 1991. This Act inserted the following clause into the Town and Country Planning Act 1990: "Where, in making any determination under the planning Acts, regard is to be had to the development plan, the determination shall be made in accordance with the plan unless material considerations indicate otherwise."

No recent Minerals Local Plans for the slate producing areas of Britain had been adopted by the end of 1993, although all were in preparation. In these areas the more general policies of the Structure Plans provide the basis on which planning applications are considered. These plans typically contain general policies encouraging the use of secondary aggregates as alternatives to primary sources, but without a full programme of future development in the area it is difficult for planning authorities to make greater provision for secondary aggregate supplies to meet local demands.

2.4.5 Planning permissions and conditions

Any application for planning permission for mineral working is considered by the mineral planning authority (development control authority in Scotland) and may be refused or granted subject to conditions. In the case of

slate working these applications may typically cover such aspects as:

- extraction of slate from an undisturbed site, a site where previous planning permissions have expired, or an extension to an existing site, with current permission;
- ancillary operations on an existing site;
- new, or extension to old, tipping areas for waste;
- reworking of material from old mineral waste deposits;
- interim uses of mineral sites, such as waste disposal;
- restoration and final use for mineral sites.

The range of matters which can be covered by conditions is discussed in some detail in MPG2 (DOE 1988b). Paragraph 55 of MPG2 states that the power to impose conditions can enable many developments to proceed where it would otherwise be necessary to refuse permission. With regard to mineral extraction, conditions serve the additional purpose of securing the environmental acceptability of proposals during and after extraction. General advice on the use of conditions is given in PPG1 and DoE/WO Circular 1/85(DOE/WO 1985). MPG2 includes a summary of matters which it may be desirable to cover by the imposition of planning conditions, including:

- time limits;
- access and protection of the public highway;
- working programme;
- environmental protection;
- surface water;
- drainage and pollution control;
- landscape works;
- restoration and aftercare.

The DOE has also issued MPG 7 (DOE 1994c) specifically to provide guidance on the restoration of mineral workings. Any conditions imposed must take into account the planning authority's policies as stated in the appropriate development plans and the responses of statutory and discretionary consultees to the application.

2.4.6 Environmental assessment

A planning application for a major slate quarrying development or a development in a sensitive area may need to be accompanied by an environmental assessment. The requirements for an environmental assessment and the guidelines for deciding when one is required are given in The Town and Country Planning (Assessment of Environmental Effects) Regulations 1988 and DOE Circular 15/88 (Welsh Office 23/88). Extractive industries and waste disposal are included in Schedule 2 of the EA Regulations and therefore the requirement for an assessment is dependent upon whether a project is likely to give rise to significant environmental effects by virtue of factors such as nature, size or location. Proposals for new slate extraction or waste tipping or the disposal of waste into old quarries may require an environmental assessment. It is increasingly the practice of planning authorities to request an environmental statement in order to ensure that relevant factors are examined during the preparation of a scheme and that the information is available to the planning authority in determining the application.

2.4.7 Reviews of planning permissions and IDO sites

Planning authorities were given a statutory duty to review mineral working sites, under the Town and Country Planning (Minerals) Act 1981. They were also given additional powers for the amendment of existing mineral permissions and for the abatement of the compensation payable as a consequence of such abatements. However there has been a general reluctance by planning authorities to use these powers in many cases since any substantial revision of conditions would require them to pay compensation to the operator.

The Planning and Compensation Act 1991 requires the holders of Interim Development Order permissions to register them and to submit schemes of operating and restoration. In March 1992 the government stated its intention "to update standards as a whole, regardless of the date when the permission was issued", and has now set out consultation proposals to achieve this (DOE 1994a).

2.5 Physical features produced by historic and modern working methods

2.5.1 Origins of workings

The majority of current slate-producing sites commenced activity in the 18th or 19th centuries. Their current operating methods reflect both modern technology and the methods used in earlier times. Modern activity is often intermixed with structures and remains of activities from previous generations.

The publication 'The Slate Industry' (Williams 1991) provides an excellent summary of working methods. The main features of the slate industry are described here under the headings of:
- slate extraction;
- slate processing;
- slate transport.

2.5.2 Slate extraction

Extraction processes developed in three broad categories according to the angle at which the slate veins were lying -the dip of the bedding plane. These categories, open quarry, pit quarry and underground quarry are shown in Figure 2.2. Each extraction method created its own waste tipping pattern, described in 2.5.9.

Open quarry This method was employed to extract slate lying relatively close to the surface of a hill or mountain.

Gallery systems consisted of horizontal benches separated by vertical intervals of 9 - 21 metres (30 - 70 feet),allowing the simultaneous working of one face by several teams of quarrymen. Galleries were first introduced by James Greenfield in 1799 at the Penrhyn Quarry, and it is there and at Dinorwig Quarry that this method is best seen.

Pit quarry Where the slate dipped almost vertically into the ground, its extraction required the opening of large pits. The most notable example of pit extraction is at Delabole, Cornwall, where the cavity is over 150 metres (500 feet) deep and about 1.6 km (1 mile) in circumference. This method was also used at Nantlle, Gwynedd and in the 'slate islands' of Easdale and Ellanabeich, Argyle. Where the topography permitted, a cutting or tunnels would be made to remove material from the pit without the need to haul it to the quarry lip, and to provide drainage without pumping. Haulage and pumping required power and led to extensive developments in water power, steam power and transport systems.

Underground quarry Where the slate dipped at a more gentle angle into the mountainside it was necessary to follow the rock underground. Thus slate mines, as opposed to slate quarries, were opened. It is thought that the technique of underground extraction was initially developed in Cumbria, where short tunnels were opened out into chambers where usable slate was found. This method, known as 'close-head' quarrying, was being followed towards the middle of the eighteenth century at Walna Scar Quarry above Seathwaite. Two men from that area, William Turner and Thomas Casson, introduced the underground methods to North Wales when they took over a quarry in the Conwy valley in 1798 and then the Diffwys Quarry at Ffestiniog in 1799. Large cavities were carved out but in order to maintain the roofs some of the slate had to be left. In the 1820s a system known as the chamber and pillar was introduced by Samuel Holland. Pillars proved to be inadequate and so a continuous line of slate - a wall - had to be retained, although the term 'chamber and pillar' continued in use. Typically, when a chamber was 21m (70ft) wide another would be started leaving an intervening slate pillar of 9m (30ft) width to support the roof. By this method 30 per cent of good workable slate remained in the ground. The largest slate mine in the world was at Gloddfa Ganol, Blaenau Ffestiniog, Gwynedd. It had 32 levels of floors, each about 12 metres (40 feet) deep, with about 70 km (43 miles) of connecting tunnels. The cliff quarries south of Tintagel, Cornwall, used a variant of the mine/underground quarry technique, whereby exposed slate was quarried and mine entrances developed directly from the cliff. These were abandoned in the early 20th century and form a distinctive feature of this section of coast.

In areas of former underground working, current practice is to remove the overlying material to extract the residual slate walls or pillars.

Figure 2.2   Slate extraction methods

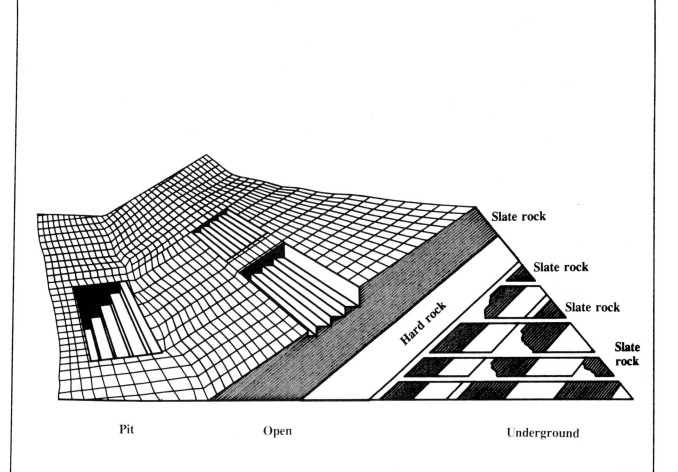

Slate quarries can be placed in three broad categories: open, pit and underground. Where the slate was dipping steeply down a mountainside, the quarry would be developed along terraces called galleries; if the veins dipped steeply into the bottom of the valley then the quarry would develop as a pit; in places where the slate dipped into the mountain side, extraction took place on floors reached by tunnels, called adits, driven in from the surface.

2.5.3   Modern extraction methods

The operating methods used at current workings are summarised in Table 2.5. Until the mid-twentieth century, slate extraction relied on manual handling of slate blocks and waste rock, and all materials had to be reduced to a size which could be lifted onto tram wagons by hand, block and tackle or winch. Black powder was used to blast the rock and to dislodge blocks of slate which would then be levered away from the face for further reduction. The process was relatively selective, and the proportion of extracted material which was tipped as waste was dependent largely on the quality of the rock and the skill of the quarrymen.

New open quarrying techniques were developed during the late 1960s and the 1970s to utilise technology and machinery developed in other extractive industries. 'High blast' techniques were introduced to dislodge and shatter massive sections of slate for removal by large wheeled shovels and dump trucks (Blunden J.1975). These techniques were financially economic but wasteful of slate rock, destroyed the old galleries, and increased the proportion of rock sent to the waste tips. More recent moves towards greater efficiency in the larger workings have led to a return to carefully designed blasting using minimum charges, exploiting natural weaknesses in the slate bed to dislodge slate blocks with the minimum of damage. Wire-sawing and chain-sawing techniques have also been introduced where the slate beds are suitable. Wire-saws consist of a loop of high-tensile wire, coated with artificial diamond beads, which is threaded through drill-holes at right-angles to each other and pulled through under tension to saw out the slate block.

The process enables quarry operators to extract rectangular blocks of 300t or more which can be sub-divided more efficiently in the production process. Slate is therefore used more efficiently and considerably less waste is generated than with blasting techniques.

Chain saws are used at Aberllefenni mine, to extract a slate which is used exclusively for sawn architectural products. The saws create very little waste, and can extract blocks precisely to the sizes required for processing.

2.5.4   Slate processing

Slate production is a rare example of an industry where extraction of the raw material and processing to the finished product often occur on the same site.

Until the large-scale modernisation of the 1960s and 1970s the processing entailed three main stages. Firstly, a block was extracted from the quarry face and reduced at that point to a manageable size for transportation. Secondly, away from the quarry face, in a safe and clear working area, the block was sawn to a size fitting as closely as possible to the finished product. Finally, the skilled splitter or 'river' set out to obtain the maximum number of slates and these were squared off or finished by the 'dresser'. In Cumbria the top end of the slate was rounded off by the 'whittler'. After the second stage, slate that would not split thinly would be split into thicker slabs.

Until the end of the eighteenth century the second and third stages were conducted in the open air and in the

Table 2.5 *Operating methods in use in 1993*

Area	Methods
Lake District/Cumbria	Open quarry Underground (very small scale)
Bethesda, Gwynedd	Open quarry
Nantlle, Gwynedd	Open quarry
Blaenau Ffestiniog, Gwynedd	Open quarry Untopping Underground (Maenofferen only)
Aberllefenni, Gwynedd	Underground
Delabole, Cornwall	Pit quarry
North Cornwall	Shallow pit excavation (c. 5m) Open quarry

crudest manner. Gradually, with improvements in conditions of work, buildings were erected. These were basic at first, but increased in sophistication during the nineteenth century particularly with increased mechanisation in the 1850s. These buildings were invariably built with stone or materials obtained from the site, including slate rejected from the splitting sheds. Their remains, in various states of decay, may be found at almost all abandoned quarries.

Mechanisation brought about the centralising of processing which led to the development of the 'slate mill'. However, this procedure was not followed universally and hence there are some regional differences in the archaeological heritage. The Ordovician slate of Ffestiniog was more amenable to mechanisation than its more brittle counterparts in the Cambrian slate regions of Wales and so the development of this centralised structure is best seen in the Ffestiniog area.

Slate mills today operate quite differently in that the internal transport system is based on the fork-lift truck rather than wagons on rails, and blocks are now sawn centrally rather than on a number of smaller saw tables. Computer control and laser guidance have been introduced to ensure that each block produces the maximum number of saleable slates. Thus the interiors of mills are quite different from the traditional pattern. Slate mills that depict the traditional pattern are disappearing because of re-use, land reclamation or deterioration. Only two or three remain in Blaenau Ffestiniog and two, with a slightly different pattern, in Dinorwig, Llanberis. There were virtually no large mills built in the Lake District until this century and whilst there are a number of interesting structures associated with the cliff quarries of Cornwall these are now in a precarious state of decay. In Scotland, visible surface remains are few and far between. In Cornwall and Scotland processing was carried out mainly in smaller shelters with mills being used for slab working only.

Modern production sites are accompanied by modern buildings housing processing, offices and storage. These range from steel-framed industrial buildings to old 'portakabins' and similar temporary structures which are out of keeping with the older slate-built buildings and the surrounding landscapes.

2.5.5 Dimension stone production

Roofing slates have always been the most valuable slate products, but slate has been used as a building material for centuries. Many producers have invested in modern stone sawing, polishing and finishing equipment to enable them to manufacture items such as sills, worktops, vanity units, gravestones and monuments from offcuts and poorer quality slate which might otherwise be tipped as a waste. Other producers concentrate on the production of dimension stone and quality finished products which have a high added value, in order to maximise the revenue from slate which is not suited to splitting.

2.5.6 Transport systems

The particular requirements for transport within slate workings led to the development of some extensive and unique features for the transportation of materials. These features and their settings are therefore of considerable importance for industrial archaeology. The transport of slate within a site falls into three broad categories:

- blocks of slate from working surface to the processing area;
- finished slate from processing area to the stockyard/outlet;
- the carriage and disposal of waste.

2.5.7 Transport from working surfaces to the processing areas

In the earlier stages of quarry development, processing areas were at the same level as the working surface, but with continuing exploitation and more centralisation, blocks of slate had to be carried considerable distances through the working and to different levels. The 'water-balance', a two-sided lift mechanism in which the load was counterbalanced by water, was introduced to raise slate from the lower levels of pit quarries. Eight water-balances were introduced into Penrhyn Quarry in 1848, but now only two remain. Chain-inclines, and 'Blondin' systems which were carriages on aerial ropeways, hauled wagons from pit and underground quarries up to the mill floors. Now the only surviving Blondin system is found in Pen yr Orsedd Quarry in Dyffryn Nantlle. An aerial ropeway was installed at the Honister Crag quarry, Cumbria, in 1927 and used until the early 1960s (Cameron, 1989). In the cliff quarries of Cornwall horse-whims, which were horse-powered windlasses, were the main haulage method used.

2.5.8 Transport of finished slates to stockyard/outlet

These yards were mainly located next to the mill, but from the beginning of the nineteenth century rail links to the ports were introduced and so rail-head stockyards were developed. One of the most distinguishing features of the quarrying landscape in Gwynedd is the incline which, in essence, was a tramroad on a slope operated by a cable wound from the drumhouse. These features are the clearest link between the quarries and their outlets, leading in many cases directly into the villages which developed around the rail heads.

2.5.9 Carriage and disposal of slate waste

Waste was always disposed of at the most convenient point, since any expense incurred in its disposal was unproductive. The commonest method was to move the waste in wagons via simple horizontal tramways and to end-tip on the hillside or land nearest to the point of origin. Waste was sometimes tipped into a worked out quarry. Larger open quarries using the gallery system developed extensive tiered waste tips each linked by a tramway to the corresponding gallery of the quarry. This system of tipping is clearly seen at many Welsh sites but was not well developed at the smaller workings elsewhere. Pit quarries required the haulage of all waste to the surface level for tipping in massive heaps alongside. Such disposal systems required considerable power supplies. At many of the more extensively developed quarries of North Wales elaborate tramway embankments and viaducts had to be built as tip complexes grew. Some of these structures remain as dramatic features but many have collapsed or been lost through regrading or new tipping. Coastal quarries in the Scottish 'slate islands' of Easdale and Ellanabeich, and the cliff quarries of north Pembroke and north Cornwall simply tipped their waste into the sea, and so no waste tips are to be seen.

All the transport of raw material in currently active slate workings is carried out by road vehicles or site dump trucks carrying as much as 50t per load. Within the processing plants, forklift trucks are generally used. At Penrhyn, conveyors are used for the disposal of off-cuts and small waste and in the production of crushed slate products. At Aberllefenni, Corris, an electric rail system is used for all underground transport. The use of wheeled vehicles dictates that haul roads are constructed with acceptable gradients and turning radii, whatever the topography. These haul roads and the safety barriers or slate blocks placed to delineate steep falls, are therefore conspicuous features of many slate workings.

2.5.10   Power sources

Power was required for haulage unless gravity systems were adequate, for slate processing and for pumping water from workings without gravity drainage.

Water power became an essential requirement in slate production as mechanisation increased. From the 1840s onwards larger concerns relied heavily on water to work the mills. However, water power was unreliable as water could be scarce in the summer and might freeze in winter. Other, more controllable sources had to be exploited although remote and lesser sites maintained their use of water. In one such site, Rhos Quarry on Moel Siabod near Capel Curig, Gwynedd, a waterwheel remained in use until the quarry closed in 1953.

Steam power became a more viable alternative during the nineteenth century, particularly where the potential production of slate warranted the investment in steam power. The first steam engine in the industry was a pump at Hafodlas Quarry, Dyffryn Nantlle, Gwynedd, installed in 1807. In 1834 a steam engine called the Speedwell was installed at Delabole.

Locally-produced hydroelectric power was first introduced to the Ffestiniog area in 1891 and led to the construction of three hydroelectric stations which today supply the National Grid. Electric power is now the primary power source for static equipment in active quarries, and so overhead cables and similar equipment are present alongside the remains of former power systems.

2.5.11   Narrow gauge railways

In Wales, the industry developed in conjunction with narrow gauge railways which were a primary mode of transport for finished slate to ports and railheads. Although slate is no longer transported by rail many of the lines survived, or have been reconstructed to operate as tourist attractions. The link with the slate industry is of considerable significance to the popularity of these lines.

2.6   The characteristics of slate waste and tips

2.6.1   Slate waste

Waste tips are the dominant visual feature of almost all slate workings, typically occupying more than half the working site. Data for the areas of land affected are given in Tables 2.3 and 2.4.

Modern production of roofing slates and dimension stone produces waste at all stages of the process, although great care is taken to maximise the saleable element at each stage. The types of waste produced are:

Overburden and 'development' rock Material which has to be removed to expose the slate vein. Typically a thin layer of upland soil will overlie weathered slate, poorly metamorphosed slate or hard igneous metamorphosed rocks. With care the soil and subsoil can be used for restoration work, and some rocks can be sold as construction materials as shown in Case Study 1. Commonly, however, virtually all the material is tipped as waste. If development rock is loosened by blasting the larger material is usually reduced in size by secondary splitting to allow its removal to tips.

Quarrying waste Irregular pieces from the slate vein, damaged by blasting or splitting or otherwise of unsuitable quality. Modern machinery can remove and tip blocks of up to 15 tonnes.

This waste material is angular, irregular and varies greatly in size. Blast design aims to minimise such wastage, and the use of wire sawing can substantially improve the percentage of usable 'block' recovered. Not all slate is amenable to such techniques, however.

Sawn ends Slate 'block' is reduced in size by large circular saws which, in the most modern operations, are laser-guided and computer-controlled to minimise wastage. The production of architectural products and dimension stone also creates sawn off-cuts. These off-cuts consist of blocks and sawn ends of up to 0.5 tonne.

Trimming waste After the reduction of blocks to the size of the roofing slates being produced, blocks are split into individual slates which are then trimmed to standard sizes. This results in fragments of thin slate sheet, reject slates and splinters.

Mill fines Mill fines are the 'sawdust' of slate processing. Sawing is water-lubricated and so the extremely fine-grained waste particles must be separated from the water before it can be recirculated or discharged. Settlement ponds, or modern filter presses, are used to produce a wet cake of clay-like slate dust which has no useful application at present and is disposed of in the tips.

2.6.2 Tipping methods

Before the introduction of modern earthmoving and materials-handling plant into slate quarrying, all waste materials had to be reduced in size for man-handling or lifting by block and tackle. The waste was then transported in tram or narrow gauge rail wagons and end-tipped to form fan-shaped tip faces. Cleavage of the waste had produced particles with a high length to thickness ratio. These particles tended to slip when tipped, to form a thin layer orientated parallel with the tip face.

The range of particle sizes deposited in waste tips increased when modern earthmoving equipment was introduced, since blocks of 10t or even 15t could be moved. When tipped, the larger blocky material tended to roll down the tip face rather than to slide, and so a blocky layer was formed at the toe of the tip. Finer material would come to rest above this layer, without a distinct plane of orientation. Subsequent vehicle loads composed entirely of platy splitting waste would slide and spread out in a layer orientated with the tip face. As the volume of waste within each load increased with the introduction of progressively larger vehicles, the uniformity of waste tips decreased.

2.6.3 The content of waste tips

An examination of tips at Penrhyn Quarry (Parkman 1991) recorded that the material sizes in tips ranged from silts to large boulders. In tips formed before the mechanisation of the 1940s, there is a predominance of material in the fine gravel to small boulder range. In tips formed in the period 1940 - 1960 there is a high proportion of cobble-sized and boulder-sized particles in a layer at the surface.

Observation of waste tips before and during excavation confirms that where a predominance of boulder-sized material is present at the tip surface, finer-grained material has migrated downwards through voids under the combined effects of water and gravity, until further movement pathways are blocked. Tips of apparently coarse waste often contain fine-grained material within one or two metres of the surface. Weathering and degradation of slate waste is an extremely slow process except where poorly-metamorphosed shales and mudstones were tipped with the true slate. Fine-grained material washed from the surface layer is not therefore replenished by weathering.

Slate particles have a Specific Gravity of 2.7-2.9, with a modal value of 2.8 (Parkman 1991). Slates formed under lower geological pressures, tested by Peacock (1983), were found to have specific gravities in the range 2.685 to 2.710. The theoretical relationship between void ratio (voids ÷ solids) and dry bulk density is given in Table 2.6.

Table 2.6 *Void ratio and dry bulk density of slate waste tips*

Void:solid ratio	Dry bulk density t/m^3
1:1	1.40
1:2	1.87
1:3	2.10
1:4	2.24
1:5	2.33
1:10	2.55
0:1	2.80

In-situ bulk density determinations at two locations in Penrhyn Quarry (Parkman 1991) showed the different characteristics of modern and old tipping methods. The water replacement method for coarse grained soils, BS 1377 Pt9 test 2.3, (Anon 1990a) was used. At location 1, modern plant trafficking had crushed and compacted a 0.2m depth, producing an in-situ density of 2.48 t/m^3. At location 2, end-tipping between 1889 and 1914 had left finer-grained waste with an in-situ density of 2.30 t/m^3. These values were at the higher end of the typical range (1.7-2.4 t/m^3) deduced by Parkman from the literature.

The extreme porosity of tips is also borne out by the experience of reclamation schemes in which tips were excavated and deposited as compacted fill. Typically 4 volumes of tip material would produce 3 volumes of compacted fill, and compaction ratios of 3 into 2 were sometimes recorded (Richards I. pc). If a finished in-situ density of 2.4 t/m^3 is assumed, it may be calculated that the density in the undisturbed tip was 1.8 t/m³ or 1.6 t/m³, *ie* between 36% and 43% voids. Because the replaced and compacted fill is less dense than the original solid rock it would be theoretically possible to re-fill many quarries with the waste tips that they generated, even if up to 10% of the original rock was sold as slate products. Improvements in the utilisation efficiency of more modern quarries mean that less waste is generated, and even if all waste is returned to the quarry hole there could be a shortfall of material.

2.7 Disposal of current production wastes

2.7.1 Waste arisings

Slate production has always involved the creation of vast waste tips. The introduction of more modern working methods, new products derived from material which previously would have been tipped, and the practice of tipping waste into old quarries, have all reduced the rate of growth of new waste tips, but even so, some modern quarries are only able to sell less than 1% of the rock which has to be moved (Hughes B. pc). Modern production sites now vary in their tipping pattern from those where virtually no material is tipped above ground to those which still tip large volumes of waste in new heaps (Box 2.1).

2.7.2 Current waste tipping

The current sites of active slate production are all in working complexes which have a long history of slate extraction. As a result the boundaries between one site and another are rarely clearly defined physically, and are complicated by amalgamation of ownership. The data in Table 2.7 indicate the scale of quarry operations and tipping at the largest sites.

Historically, flat-topped waste heaps provided the only sites available for processing buildings, storage and transhipment facilities. Modern quarry operations are rarely constrained by lack of such space, and so have no immediate need to tip the waste in a coordinated way to facilitate further development. As a result, at sites where waste tips are not controlled by modern planning and restoration conditions surface waste tipping is not being conducted towards a planned future land use or restoration plan. These sites form the great majority of slate working sites. Only one of the active sites visited had either a defined new use for tipping sites or a plan for the ultimate shape and revegetation of these tips, and in this case it was dictated by location. In the majority of cases the existing permitted tipping facilities are sufficient to meet requirements for many years. However, the cost of transporting slate waste to remoter permitted tipping sites is prompting operators to seek permission for disposal sites nearer the source of extraction.

Box 2.1   *The current pattern of waste disposal*

Cumbria. Waste is primarily tipped in surface waste heaps. At Petts Quarry all waste from extraction and processing is tipped on site. Some tipping is planned to fill voids within the workings.

At Elterwater, Brandy Crag and Moss Rigg, waste from extraction is tipped near the quarries. Usable material is transported to Kirkby in Furness for processing and so process waste is tipped there in a large tip together with waste from that site.

Penrhyn Quarry, Bethesda. Waste tipping within the quarry bowl has progressively replaced surface tipping. Since 1980 only soil and soily overburden have been tipped in surface waste heaps, to encourage regeneration.

Pen yr Orsedd, Nantlle. Waste is tipped within previously quarried areas.

Blaenau Ffestiniog. The majority of waste is tipped in surface waste heaps. The slate reserve dips almost vertically and so tipping within some quarries would prevent future extraction. The backfilling of quarries was common practice at Llechwedd.

Aberllefenni slate quarry, Corris. All extraction is from underground workings and so no waste is brought out. Processing waste is minimal as all raw material is sawn to order. Surface waste tipping is now just 100t/year.

Delabole slate quarry, north Cornwall. Overburden and unusable rock is tipped within the quarry void. All material brought to the surface is processed and sold in some form.

Small workings, north Cornwall. Waste from these shallow workings is tipped in small waste heaps or back into worked-out areas.

Table 2.7 *The scale of current slate workings*

Site	Area of excavation hectares	Area of tipping hectares	Total[2] hectares	Total of planning permisions[3]
Cumbria				
Kirkby	10[1]	65[1]	75	193 (IDO)
Brandy Crag	n/a	n/a	0.4	1.7
Broughton Moor	n/a	n/a	2.5	74
Bursting Stone	n/a	n/a	1.8	4.8
Elterwater	n/a	n/a	6.8	15.4
Moss Rigg	n/a	n/a	1.2	6.0
Spout Crag	n/a	n/a	2.0	4.9
Petts	n/a	n/a	2.4	6.9
Gwynedd				
Penrhyn	108.6	130.2	238.8	240-250 (IDO)
Ffestiniog/Gloddfa Ganol	13.4	96.5	109.9	170.3
Cwt y Bugail	4.4	1.6	6.0	6
Cwmorthin	6.3	27	33.3	60.7 (6.0 ug)
Croes y Ddwy Afon	n/a	n/a	10 approx	n/a
Llechwedd) Maenofferen) Diffwys-Fotty)	45.8	114.8	160.6	678 (358ug)
Manod-Graig Ddu	6.5	44.5	51	57
Pen yr Orsedd	33	25	58	n/a
Aberllefenni	3.8	16.8	20.6	20.6 (12.6 ug)
Cornwall				
Delabole	14.8	39.1	53.9	104

Notes:
[1] Estimated from site visits or maps
[2] Data supplied by Cumbria County Council, Gwynedd County Council, Cornwall County Council, 1988 Mineral Surveys
[3] Underground mining permissions shown 'ug'
n/a Not available

2.8   Tip and quarry stability

2.8.1   Regulation and legislation

The design and management of tips and related structures is regulated by The Mines and Quarries (Tips) Act 1969 and The Mines and Quarries (Tips) Regulations 1971. This legislation is enforced by the Inspectorate of Quarries, a part of the Health and Safety Executive, where 'active' or 'closed' tips form a part of working quarries. The safety of the public in relation to disused, classifiedtips under Part II of the Act is a function of the local authority. The Act and Regulations in general define the responsibilities of operators and statutory bodies, and require that 'competent persons' are appointed to design and inspect tips. A useful summary of this legislation is provided in 'Handbook on the Design of Tips and Related Structures' (Geoffrey Walton Practice 1991).

The Regulations require that tip stability, tipping operations and the proper functioning of drainage systems are inspected at set minimum intervals according to the status (active or closed) and class ('Classified' or not) of the tip. These records must then be kept for inspection. When quarry operations cease the tip may, on notification to the Inspectorate, become a 'disused' tip.

Site inspections by HM Inspectorate of Quarries are made typically 2-4 times per year. The largest operators are regarded as industry leaders and are encouraged/expected to lead the way in setting operating standards. These operators therefore attract visits more frequently than the small operators.

2.8.2   Tip stability

When interviewed as part of this research, operators confirmed that slate waste is inherently stable and free-draining once tipped and that no instances of tip movement had been recorded (quarry operators, pc). Any such occurrence would have to be reported to the Inspectorate of Quarries. No other records or recollections of the collapse of slate waste tips were found during this review. In one detailed study of tip stability (Parkman 1991) observations and topographic surveys were reported to have shown that loosely-tipped slate waste comes to rest at approximately 40-42°, the precise angle in each case being determined by the particle size and shape. Surveys at Penrhyn Quarry recorded an average tip surface inclination of 36°. Older tips were measured at 34°, and locally steep sections of 40°-45° were measured but these were recorded as unstable. Material being tipped from a conveyor was observed to come to rest at between 30° (platy shapes) and 38° (blocky shapes). Material resting at over 32° was easily mobilised by disturbance. The topographical survey of the Fotty tip, Blaenau Ffestiniog, before regrading recorded a surface angle of 45° over much of the tip. The waste consisted predominantly of larger blocks of material (Byrne C. pc).

2.8.3   Tip disturbance

Slate waste at or near its angle of repose is easily mobilised by disturbance since there is no cohesion between particles and the degree of interlocking is generally very low. Tips formed by end-tipping, as described in section 2.6.2, will experience surface sliding of particles dislodged by livestock, humans, freeze-thaw or thermal expansion/contraction. Tips formed of blocky material or fines with no uniformity of particle orientation can be excavated to give temporarily stable faces which are near vertical, but these are vulnerable to any disturbance. Such over-excavation of the toe of tips was observed at some sites where waste has been extracted for use or to create additional working space (Williamson 1988). The removal or regrading of large tips by excavation at the toe should be carefully planned and controlled.

The slippage of unrestrained slate waste after disturbance is very localised unless the largest fragments are mobilised, since the kinetic energy becomes dissipated and insufficient to mobilise further material. Over many decades, the tip surface will tend to 'shed' the least stable particles so that further movement requires greater disturbance. In general the slippage of material presents a problem only for property or structures immediately adjacent to the tip face. Tips have stood at the rear boundaries of houses or at the sides of roads for many years without serious incident although the potential for accidents cannot be ruled out. Slate waste slipping onto a highway could, for example, lead to injury or damage, and waste slippage could interfere with surface water drainage. Surface instability is a major inhibitor of the natural colonisation of waste tips and so stabilisation measures may also be required as part of revegetation works.

2.8.4   Tip drainage

The ratio of voids to solids, and the large void dimensions found in slate waste tips, ensure that tips are free-draining. The tendency of large blocks to roll preferentially to form a coarser layer at the base of tips also ensures adequate lateral drainage over horizontal, relatively impervious ground. Tips do not therefore alter the groundwater or surface water regimes to a significant degree, and no problems of impeded drainage, or tip instability, have been recorded. The need to tip waste near the site of production, and the lack of available land in many locations, led to the construction of tips directly over small surface water-courses, without the construction of culverts. This practice would not be permitted today. The investigation at Penrhyn Quarry (Parkman 1991) recorded that no groundwater was found within the tips despite records of a stream having been covered. The tip material was found to be 'slightly damp', or locally 'damp' where silty deposits were located. This situation contrasts markedly with tips of cohesive material or easily weathered colliery shales which degrade and can become water retentive, leading to instability. The tolerance of slate waste tips to water was demonstrated in Blaenau Ffestiniog in 1980 when a storm, at least as severe as a 1 in 100 year event, led to mountain and quarry run-off water passing into slate waste tips and issuing in great volumes from the tip sides. No movement of the tips was detected (Richards I. pc).

2.8.5   Ground stability

Many hillside tips rest on ground of considerable gradient, and yet no records of failure due to slippage of this ground are known. Quarries in upland areas began where slate rock was at or near the surface and overburden was thin or absent. It is assumed that the combination of surcharging and free drainage is sufficient to prevent slip failures.

2.8.6   Retaining walls

Many tip complexes contain retaining walls, tipping heads and similar structures built from unmortared slate. These structures deteriorate progressively and collapse, unless maintained. Tips retained by dry slate walls have led to concern where collapse would endanger the public. For this reason, reclamation works have been carried out at Abercwmeiddaw Quarry, Upper Corris, and a repair scheme was implemented at Vivian Quarry, Llanberis. Walls have been removed from sites at Tan yr Allt, and Ffrancon View, Bethesda. Retaining walls showing severe cracking were observed at a number of sites during this research. The collapse of tipping heads can cause surface slippage of material.

2.8.7   Quarry faces

Abandoned quarry faces, in common with other hard rock exposures, are subject to natural mechanisms of deterioration although slate's inherent resistance to weathering and degradation makes it a relatively stable material. Slate veins, and the adjacent rocks, possess natural faults and weaknesses together with any weaknesses such as overhangs left at the cessation of quarrying. These weaknesses can result in quite major failures. At Penrhyn Quarry a known area of faulted weakness slipped in November 1989, after heavy rainfall. Monitoring had warned the quarry operators and the area had been cleared on the previous day. The landslip involved 3,200,000t of material, the largest block weighing about 3,500t. At Twll Ballast in the Nantlle valley complex, a smaller landslip in 1982 involved a section of the quarry side which is relatively inaccessible to the public. These collapses do indicate the need for continued vigilance at any site to which the public have access. Guidelines on the hydrogeology and stability of excavated slopes and quarry faces are given in the Handbook on the Design of Tips and Related Structures (Geoffrey Walton Practice 1991).

2.8.8 Underground stability

Underground slate workings, such as the 'close-heads' of the Lake District or the more complex adit and 'chamber and pillar' workings of Honister or Blaenau Ffestiniog, are noted for the collapses which occurred during or after extraction (McFadzean 1986). In addition to the risks of roof collapse, underground workings present hazards from shafts and decayed temporary structures such as timber staging.

2.9 Wildlife value

2.9.1 Colonisation of waste tips

Slate waste tips are generally much more resistant to natural colonisation by vegetation than are other forms of mineral extraction waste. The characteristics which restrict the establishment of plants are principally physical:

- coarse particle size promoting free drainage;
- lack of soil sized particles to retain water or protect germinating seed against desiccation by wind and sun;
- extremes of surface temperature;
- elevated and/or exposed locations;
- surface instability causing damage to young plants;
- lack of protection from rabbits and sheep.

These characteristics are exacerbated by nutrient deficiency due to leaching from the waste which is inert and lacks ion-exchange capacity. Abandoned slate tips do not therefore have a characteristic vegetation as do former limestone quarries or metalliferous mines. Nevertheless former slate workings exhibit some wildlife interest. The tips themselves rarely support a closed cover of herbaceous vegetation but voids between plates of slate may become colonised by relatively uncommon plants such as Parsley Fern (Cryptogramma crispa) or the more common Foxgloves and other acid tolerant herbs. Where there is little or no grazing, Hawthorn, Rowan and Oak may establish and slate heaps have become colonised to form open woodland (Ratcliffe 1974). The disused workings at Hodge Close (Case Study 6) now support many self-sown birch trees on areas of finer-grained waste. Occasionally slates such as those at Honister Crag, Cumbria are slightly

calcareous and where this is so, walls and weathered material can become colonised by more uncommon plant species. Slate workings within the shelter of valleys, and particularly within the damp environment of broadleaved woodland, have been found to develop quite extensive covers of mosses and woodland ferns.

Older slate heaps can support substantial colonies of lichens but lichens take many years to colonise freshly exposed slate, and colonise cracks and pits more readily than exposed faces (Armstrong 1980). The principal lichens which colonise slate are, however, the more common species and slate is not known for particular lichenological interest.

2.9.2 Habitats for wildlife

Former industrial buildings, caves and holes are important habitats for bats and birds at many abandoned slate workings. Some slate caves support nationally rare Horseshoe Bats and Choughs, whilst other birds, particularly Peregrines, use slate sites for breeding. Slate quarries can also provide important habitats for invertebrates, providing south facing slopes on which invertebrates can bask and plenty of dark moist spaces for them to hide. Common small mammals such as rabbits and rats inhabit slate waste tips and quarries, attracting predators such as foxes.

Conservation organisations do not publicise the locations of rare species, but the following details from Meirionnydd indicate the value of quarries (RSPB pc):

- Peregrines: 3 nest sites in non-working quarries. District total 46;
- Raven: 17 nest sites in quarries, of which 3 are active. District total 90;
- Chough: 7 nest sites in quarries, of which 1 is active. District total 10;
- Wheatear, Ring Ousel, Barn Owl all frequently associated with inactive quarry sites.

2.9.3 Surrounding habitats

The slate-producing upland areas of Britain are dominated by acidic base-poor rocks. The surrounding semi-natural vegetation is predominantly grassland, heathland or bracken which is often widespread in the locality, unless it has been replaced by commercial forestry. Such vegetation is usually species poor. Although the extent and openness of these types of vegetation provide valuable habitats for wildlife such as birds of prey, grouse and perhaps deer, the effect of slate extraction on them has often been minimal because the area of working is small compared to the extent of the wildlife resource. Increased pressure on semi-natural habitats throughout Britain has, however, led to an increased perception of the value of upland semi-natural habitats.

In some locations slate production has resulted in a diversification of habitat after mineral extraction has ceased by providing, for example, habitats inaccessible to sheep, caves suitable for bats and ponds in quarry holes.

Slate quarrying can have a local ecological impact where locally uncommon or important habitat has developed. Ratcliffe (1974) cites the Dinorwig quarry in North Wales as having cut into 'hanging' sessile oakwood resulting both in the loss of part of it and in enhancing its ecological value by forming a barrier against sheep with the result that the remaining woodland was the only ungrazed example of its type in the region.

2.9.4   Conflicts with wildlife

Where the wildlife resource is of particular importance then conflict may arise. An example of this is Kirkby Moor SSSI in Cumbria where 103 ha of the 781 ha SSSI is affected by an Interim Development Order permission for the extraction of slate. If extraction were to go ahead the loss of wildlife conservation value would be significant. Historically, impact on wildlife has not been a primary issue in the planning of slate extraction. However recent legislation and guidance from central government has resulted in new quarries requiring an assessment of environmental impact including appropriate consideration of wildlife matters during the planning process (see 2.4.7). Options for dealing with wildlife habitat which is threatened by slate extraction may include:

* changing the mine plan so as to work around areas of importance;
* establishment of areas of similar wildlife value within the land holding;
* management of existing vegetation to enhance its wildlife value to compensate for the habitat lost;
* transplanting of vegetation of value to a new site before slate extraction.

The feasibility of these options would be dependent on the type of habitat to be lost and the practicality and cost effectiveness of the approach. Where protected species such as badgers, bats and newts are involved then guidance should be sought from the appropriate conservation body - English Nature, Countryside Council for Wales and Scottish Natural Heritage. It is good practice to consult these bodies early in the planning process as they may be able to offer advice as to the wildlife status of a proposed site of slate extraction. The assessment of wildlife value is described in section 4.5.

The reopening of old quarries for slate production may also disturb areas of wildlife value. Of the protected species noted above bats are perhaps the most common to be encountered in former slate workings but such workings may also be of ornithological or botanical interest and some are Sites of Special Scientific Interest. Again, consultation with the appropriate statutory wildlife conservation bodies would be necessary in planning the development of such sites.

2.10   Associated abandoned land

2.10.1   Undisturbed land

Almost all slate working sites contain areas of land which are more or less undisturbed. These areas may lie between tips, between the site boundary and tips or excavations, or around old buildings. This land often supports the remains of the vegetation type which formerly covered the land before quarrying. In most cases this vegetation is a short grass sward grazed by sheep or rabbits, or is broadleaf woodland. The quarrying activity and subsequent abandonment to dereliction has frequently ensured freedom from disturbance for this land, allowing a diverse flora to develop. Consequently these undisturbed areas of land may be valuable examples of the local 'natural' vegetation, just as the quarried areas provide valuable 'artificial' habitats.

Undisturbed land can also provide wildlife with links between habitats existing on either side of extensive workings.

2.10.2   Problems of abandoned land

Unused spaces attract problem activities such as rubbish dumping, fly-tipping, unauthorised camping, motorcycling and the like. Slate workings are no exception to this, although many are too remote or inaccessible to suffer greatly. A 'rave party' held without permission at the Dorothea Quarry complex attracted over 150 people (Daily Post 1993). Many people, including farmers and landowners, regard old quarries as the obvious location to dump old cars, household goods, scrap materials and so on. This contravenes the Environmental Protection Act 1990 but many examples of such tipping were seen during the course of this study. Fly-tipped waste adds greatly to the appearance of dereliction and degrades the environment of those living nearby, thus creating or reinforcing negative perceptions of abandoned slate workings. The tipping of waste soil by authorised or unauthorised persons, frequently causes the introduction of invasive weeds such as Japanese Knotweed (*Fallopia japonica*). This vigorous perennial produces dense stands of 2-3m height which die back each winter leaving an unsightly brown mass of dead stems, adding to the derelict appearance of such sites. Japanese Knotweed was seen at abandoned slate workings in Cornwall and North Wales. The abandoned Prince Llywelyn Quarry in the Lledr Valley near Dolwyddelan is believed to have been the source of Japanese Knotweed invasion down this river valley (Palmer J. pc).

2.11 Industrial archaeology and historical heritage

2.11.1 The effect of current slate working

The industrial remains and historical interest of slate working sites were described in section 2.5. It is unavoidable in an extractive industry that current working will destroy the evidence of previous activity. This is clearly seen in the expansion of the Penrhyn Quarry, Bethesda where previously abandoned parts of the quarry are being reworked. Two water-balance inclines survive but the 'traditional' mills and transport systems have largely disappeared together with much of the old 'gallery' working system. Modern working at Maenofferen, Blaenau Ffestiniog has destroyed the internal rail system of the quarry (Davies K. pc). A similar situation exists in Delabole Quarry, Cornwall, where little of historical interest now remains.

Nevertheless, a surprising amount of interest can survive on active sites. Because of the nature of the operation, once the accessible rock has been worked out archaeological remains can survive remarkably unscathed. Diffwys Quarry, Blaenau Ffestiniog displays this characteristic quite well. This quarry has been working more or less continually since 1760 but until the late 1980s contained archaeological evidence from at least the first decade of the nineteenth century. The historical development of extraction, processing and transportation could be interpreted from field evidence. Recent exploratory work is threatening this continuity at Diffwys and at a number of Blaenau Ffestiniog quarries as the constant search for usable rock means a re-working of the 'historical' areas. Since 1989 much re-working has occurred in Cwmorthin Quarry, Blaenau Ffestiniog. Structures lost since then include a unique powder magazine as well as less valuable features that contribute substantially to the overall understanding and appreciation of the site.

Limited working in the Lakeland quarries ensures the survival of the few remains on these sites. The Coniston quarries still retain many interesting features, but the question of re-working remains over the Honister Pass quarries which have much archaeological interest.

2.11.2 Conservation at operational quarries

The current practice of 'untopping' former underground workings to remove the residual pillars of slate, the tipping of current wastes, the natural decay of old structures, the use of quarries for waste disposal and other activities have all contributed to the progressive disappearance of traditional slate quarrying patterns. Since 1980, interested parties have suggested that the sites where these patterns are best illustrated need to be identified and protected (Roberts D. pc). The mechanisms for the protection of abandoned sites, and examples, are given in section 4.7. Working sites for which planning permission already

exists have no obligation to evaluate or conserve features of interest, but one operator does have a policy of conserving artifacts by removal and subsequent refurbishment as part of a tourist attraction (Roberts W. pc) and another company works closely with the National Museum of Wales in refurbishing and safeguarding features of interest. This company is establishing a programme of recording and protection with the county archaeological trust (Law C. pc).

2.11.3 Conservation of associated features

Water wheels of 15 m diameter were commonly used to power machinery, and so where slate mills or their remains survive the associated wheel pit is often a striking feature. The provision of consistent supplies of water for these wheels often required extensive systems of leats, reservoirs and launders, which can be traced today. These features are all of interest to industrial archaeologists and the general public. The study 'Rhosydd Slate Quarry' (Lewis and Denton 1974) of the site on the Moelwyn mountain in Ffestiniog describes one example of a water power system. Sites that have not been researched may well have attached to them fascinating networks of water systems that could be used as interpretation material for educational purposes. Steam power and earlier electrical power systems such as those described in section 2.5.10, may also merit preservation.

The slate waste tips are the most obvious remains at almost all sites. These should not be dismissed simply as eyesores as they form an integral part of the heritage. There is a clear distinction between quarry waste and mill waste, and so the development of the quarry can be traced through the study of the tip pattern and the waste material, which may contain evidence of the different processing methods used.

Transport features such as inclines are the clearest link between quarry and outlet and in many cases lead directly into a village. This is clear evidence of the link between quarry and settlement - the work-place and home. In the reclamation scheme at Fotty tip, Blaenau Ffestiniog a case was made to retain the pattern of the main incline which led into the former station yard in the town. It is important to realise the significance of these links so that the quarries are not seen in isolation.

2.12 Slate workings in the landscape

2.12.1 The location of slate workings

The slate workings of Britain are all within predominantly rural areas of recognised high landscape quality. Figure 2.3 shows the working centres in relation to the North Cornwall Heritage Coast, the Pembrokeshire Coast National Park, Snowdonia National Park, the Lake District National Park, and the Queen Elizabeth Forest Park. Those in England and Wales lie within the areas

Figure 2.3 Slate workings and designated landscapes

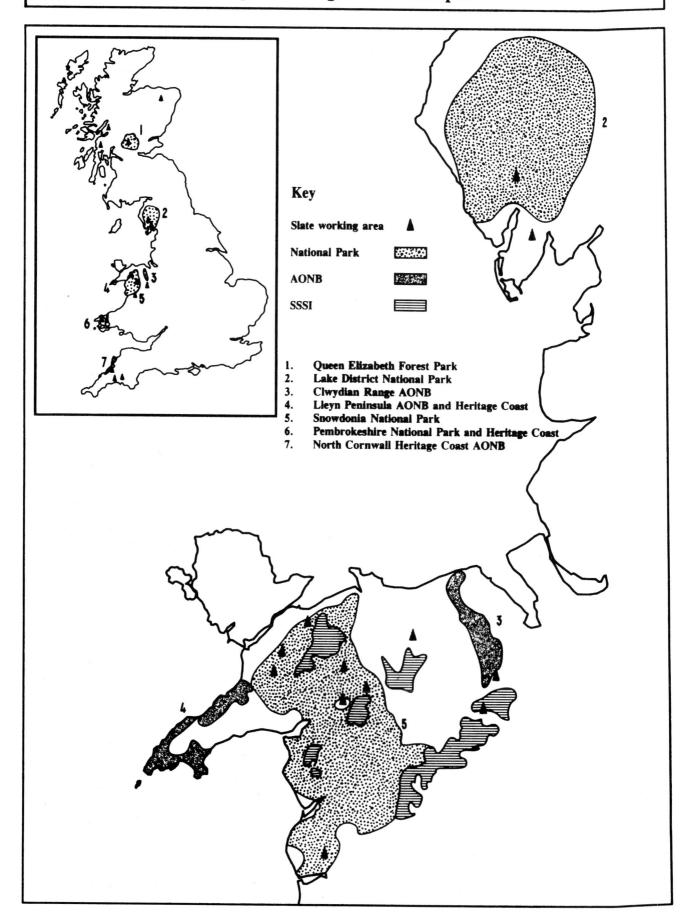

Key

Slate working area ▲

National Park

AONB

SSSI

1. Queen Elizabeth Forest Park
2. Lake District National Park
3. Clwydian Range AONB
4. Lleyn Peninsula AONB and Heritage Coast
5. Snowdonia National Park
6. Pembrokeshire National Park and Heritage Coast
7. North Cornwall Heritage Coast AONB

regarded as the 'uplands' by the Countryside Commission (1984), broadly defined as the land predominantly above 240m together with the adjoining settlements. There are certain characteristics which are common to all the uplands. Height, topography and climate combine to create a rugged terrain, wild dramatic landscapes and a harsh environment, the severity of which increases with altitude and exposure. Soils are invariably poor, while much of the vegetation is semi-natural. Rainfall is higher and temperatures lower than the national average. The uplands are sparsely populated and much less intensively used than lowland areas, and they tend to be remote from major centres of population so are very much at the margins of the nation's economic and social life.

These upland areas constitute a major part of the nation's scenic, historic, natural and recreational heritage. The uplands are of exceptional archaeological value because, in contrast to most lowland areas, much evidence of our history has survived virtually undisturbed until recent times. Moreover, many of the Nature Conservancy Council's Sites of Special Scientific Interest (SSSIs) in England and Wales are in upland areas. These qualities, together with the characteristic wildness of the upland environment and the distinctive culture and lifestyle of those who live and work there, attract large numbers of visitors annually. The uplands thus represent a major natural resource for tourism and recreation.

The Scottish quarry areas, both upland and coastal, lie in landscapes which share many of these characteristics. All have in common the fact that quarrying was a major element in the influence which man has had, and continues to have, on the surrounding landscape. The term 'cultural landscapes' is used in this report to represent the collective physical and social influences of the slate working industry and its communities on these areas.

2.12.2   Prominence

Slate workings vary greatly in prominence from those which are perceptible only to the keen observer to those which dominate the view. The degree of prominence of a slate working in a view can be assessed by considering the characteristics described in Box 2.2, and comparing the workings with the surrounding landscape. It should be noted that the colour, shape and slope of a tip contribute in varying proportions to its degree of visual intrusion. These proportions vary according to setting, material, weather conditions, season and age. Slate workings become prominent when they appear to contrast with the landscape in one or more of these characteristics.

The degree of prominence of slate workings due to these characteristics is modified by the degree to which revegetation or weathering have mellowed or softened colours, textures and outlines. The role of vegetation in obscuring and filtering views is clearly visible at Tyn y Bryn quarry near Dolwyddelan. Lichens and mosses, and weathering

combine to introduce subtle variation to the colour and texture of slate waste or rock faces, reducing the contrast between waste and natural features nearby. Slate waste in the Lake District is noticeably mellowed by weathering and lichen growth.

These characteristics should be analysed when works of reclamation or visual improvements are planned, since it is essential to identify the reasons for a site's prominence in order to design measures which will reduce that prominence.

Once the prominence of a site has been assessed the effect of that prominence can be evaluated in terms of:
- where the site is seen from, eg dwellings, roads, footpaths;
- who the site is seen by, eg residents, visitors, and their number;
- whether the site is part of a well-known or valued scene;
- whether the site is seen at close range or from a distance;
- whether the site creates an initial or lasting impression of the area.

This evaluation will help in the overall assessment of a slate working site and the need for amelioration or more drastic reclamation works.

2.12.3   Perception and conflicting opinions

Culture and pre-conceived expectations or understanding have a strong influence on the perception of slate workings in the landscape. The varying expectations of those who live and work in the area, and those who make brief visits, can lead to conflicting opinions of slate workings. A selection of consultation responses and published views is presented in Box 2.3 to illustrate this.

2.13   Public perception of slate workings

2.13.1   Historical dimensions

Slate quarries have long been a source of wonder for tourists. The 'first' travellers of the late eighteenth century always noted the presence of slate quarries when they visited the relevant areas. Descriptions are found in the works of the most famous of these traveller/writers such as Thomas Pennant and Anthony Aiken. Artists and illustrators of this period were as fascinated as writers. In the 1790s Lord Penrhyn organised a tourist programme that included a ride up from Bangor to the foot of Snowdon and on return a stop at his quarry where special pagoda-like buildings were erected to view the workings. Of course, these tourists were people of the same social class as their host.

By the mid-1850s tourism was made more widely available by the development of railways. Tourist guides,

Box 2.2 *Landscape characteristics of slate workings*

The following characteristics contribute to the prominence of slate workings in the landscape. The numbers refer to Figure 2.4.

Tip and quarry shapes   *1,2*

Slate waste tips are generally angular, flat-topped and very steep-sided (38-42°). Open quarries typically have near vertical or benched sides and a flat base. Where these workings break the skyline of a view or contrast in colour with their background, these unnatural profiles are seen particularly strongly and may interrupt the natural sweep or flowing lines of the view. In contrast, tips sited high on a hillside and viewed face-on, may be distinctive only in the outline of the tip 'footprint'. The natural landscapes of slate-working areas are rarely strongly angular or flat-topped, and where steep mountainsides exist these are usually above the areas of slate waste tipping.

Size of workings and scale of setting   *3,4*

Workings viewed from a distance within the broader sweeps of an upland landscape are generally seen as fussy or cluttered because relatively small features are seen more clearly than the small-scale features of the natural landscape (see Texture). In contrast, a slate working or tip seen close-to will often dominate the view because it is much larger than the adjacent landscape features such as trees, walls and hedges. Many large-scale reclamation schemes remain prominent because the scale of the landform and spaces created is larger than that of the surrounding landscape.

Colour and contrast

Slate waste may be indistinguishable from the natural scree, as at Honister, or may resemble exposed rock or soil as with the rustic slate of Cornwall. In most cases however the blue-grey or purple-grey colour of slate waste contrasts strongly with the semi-natural or agricultural vegetation of the surroundings. Freshly tipped waste is usually paler in colour and so more prominent. Colour is strongly influenced by weather and light conditions which can pick out subtle colour variations or render the whole view a dull grey.

Texture   *5*

Slate wastes vary greatly in unit size and therefore texture. The texture of the waste compared with the background greatly increases or reduces a tip's prominence. Tips made up of large particles tend to have a broken appearance, with light reflecting off large flat faces of slate which are interspersed with equally large areas of shadow. Conversely, finer tips have a more even appearance, often being more obvious than the coarser areas because they are seen as an unbroken block of uniform texture. The texture of the landscape is also strongly influenced by weather and light conditions which can enhance detail or blanket texture almost completely.

Buildings and plant   *6,7*

Modern quarry buildings and processing plant are alien to rural settings, in terms of their shapes, sizes, colours and materials. Most quarries have abandoned equipment, materials or scrap scattered about, and items such as cars or plant in corporate colours/high visibility colourings which are absent from the rural surroundings. Traditionally, quarry workers used the slate to build in the vernacular style which matched the houses and farm buildings of the area. The resulting buildings or ruins may be eye-catching features but are made less obtrusive by appearing to belong to their setting.

Figure 2.4 Slate workings in the landscape, refer to Box 2.2

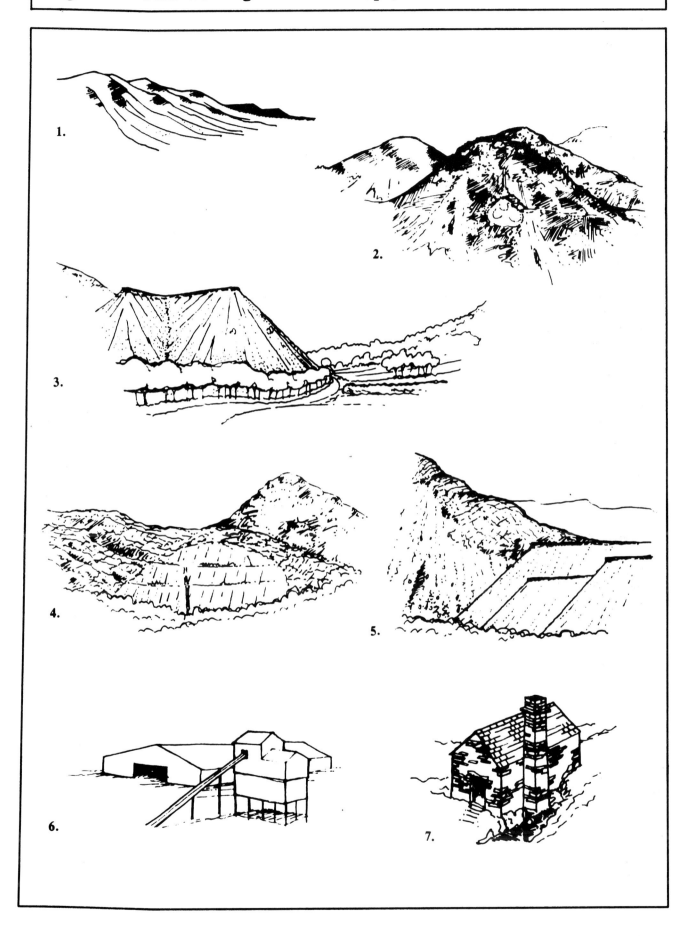

Box 2.3   *Conflicting opinions of slate workings in the landscape*

The Lake District

'Lakeland Landscape' Clarke and Harding-Thompson (1938)
"At Honister, Coniston and Elterwater the old and existing slate workings are plainly visible from a distance; but it cannot be said that these legitimate and necessary activities do much to mar the beauty of the countryside as the stone extracted, and even the quarry tips, are of the same colour as the surrounding rock outcrops." (reported by Stephens Associates 1988).

'Land of the Lakes' Bragg (1983)
"There are those who consider these to be scars on the landscape. I can see the point of that argument, but (especially since finding out something about geology) I enjoy them as specimen centres, open laboratories, in some cases open displays of the rocks we live on." (reported by Stephens Associates 1988).

'Slate Quarrying in the Lake District' Stephens Associates (1988)
This report recognises that "small, traditional quarries, formerly worked by hand, are not intrusive but provide landscape features". However, the report also refers to large modern quarries which "disfigure the landscape and require associated service roads, traffic and buildings". This contrast between traditional quarries and modern workings is an important point.

The Snowdonia National Park

This was established with boundaries determined by the National Parks Committee who reported in July 1947:-

"The quarrying of vast amounts of slate presents the most formidable problem of the North Wales National Park. Many of the waste tips are enormous and may consist of lumps of rock averaging the size of a coffin, or mounds and cascades of splintery scree. Their colour is particularly dismal and the complete absence of soil or friable material in their composition makes surface restoration an impracticable or exorbitantly costly undertaking. We have drawn the north-west boundary of the land to exclude the whole belt of exploitation through Bethesda, Llanberis and Nantlle. The even worse disfigurement from the Blaenau Ffestiniog quarries and the unattractive urban development among them, seemed to present such an extensive and intractable problem that the whole area has been excluded from the Park as an island for which it seemed inadvisable that the National Parks Commission should take responsibility. Other and smaller quarries scar the mountain sides and choke the ravines with their spoil tips in various parts of the Park but the areas excluded contain most of the worst disfigurement from the workings and buildings and room for any necessary extension of either" (reported by Crompton, 1967).

The value of selected abandoned workings as a tourist and historic resource was recognised in Crompton's 1967 paper.

"Indeed there is a minor school of thought which considers that these slate heaps are a witness to man's industry and, as such, should be left undisturbed. To leave all of these workings would be wrong but there are, however, points in favour of the preservation of one or two well located examples of disused quarries, which could in the future become of tourist and historic value."

Recent views

The development of attitude from condemnation of slate workings to a desire to conserve their place in, and influence on, the landscape has continued to the point where the Countryside Council for Wales gave as their view:

"Slate waste tips contribute greatly to the character of local landscapes, provide good examples of industrial heritage and can provide habitats for wildlife. The landscape of Snowdonia National Park, for example, is littered with the remains of numerous disused slate workings which contribute greatly to its landscape character. The value of workings varies. Much depends on the local situation in which industrial workings are situated and should be assessed according to local value. CCW would be concerned to ensure that any guidance on the reclamation/rehabilitation of sites requires the landscape role and value of such workings to be assessed".

The Council emphasised the need for modern workings to incorporate progressive restoration conditions and working practices which are sympathetic to the landscape (Davies K. pc).

such as Blacks' guide, also directed people to the quarries and urged them to marvel at the wonders of human engineering. Victorian travellers to the Lake District climbed the Honister pass to gaze at the 1000 ft. buttresses quarried for slate (Cameron 1989).

Gradually, as the horrors of the Industrial Revolution took over the psyche of writers and commentators, attitudes changed and attention was drawn to the depredation caused by industry. Thus, by the 1890s writers such as A J Bradley in his 'Highways and Byways of North Wales' lamented the scars of the Penrhyn Quarry and, in a manner which may now seem condescending, pitied the lives of the poor inhabitants of the villages.

2.13.2 'Conservationist' perception

During this century there has been a growing awareness of the 'beautiful countryside' culminating with the establishment of National Parks through the National Parks and Access to the Countryside Act, 1949. The realisation of the concept of 'protected areas' in Britain has raised many issues and highlighted many contradictions. National Parks in many people's minds should follow the American model, the 'untouched by human hand' idea, whereas in fact they are nothing of the kind. They are not 'National Parks' in the sense used throughout the world apart from Britain, but landscapes where the interaction of man and environment is at the very heart of the personality of each area. How to cope with the inherent conflict of maintaining economic viability whilst ensuring the conservation of nature and landscape is the core of the problem for National Park authorities. But this is also the core of the problem for people's perception of slate quarries, both working and abandoned, as many of the areas of slate happen to be in or close to National Parks namely Snowdonia, Pembroke and the Lake District.

A report commissioned by the Friends of the Lake District (Stephens Associates 1988) states strongly the view that all new slate quarrying, including extensions to workings, in the National Park should be subject to "the most rigorous assessment" in line with MPG 1 and "should not be allowed .. unless there is a proven special need." It also recommends that "a restoration programme for slate waste tips and closed quarries should be established as a priority." Although the report mentions the use of local slate for traditional construction, it makes no mention of the fact that many of the cultural characteristics of the area derive directly from this basic industry in terms of villages, communications and economic viability. It also highlights the perception that country areas are areas of little activity which the urban folk can, to use the phrase in the legislation, 'resort to' in order to 'get away from it all'.

2.13.3 Local perceptions

People who live in these areas have a different perception. To many, slate quarries represent a way of life - an essential part of the make-up of their society. In the slate tips and ruins around them lie the memories of the work, the joy and suffering that has been lives of past generations. Local people are fully aware of the fact that it is these things that have made them what they are. However, because the quarries represent both 'good and evil', attitudes can be ambivalent. Local people can say that all vestiges of the industry should be removed so that they can look to the future without the remains of the past around their necks. Nevertheless, they still appreciate the significance of the slate legacy.

2.13.4 Visitors today

Little research has been done to evaluate people's perception of slate quarries and tips today. Attitude surveys by the National Park Authorities in both the Lake District and Snowdonia did not specifically ask for people's opinion on slate. However, in the Lake District Visitor Survey (Anon 1987) no-one mentioned the slate industry as a drawback and surveys in Snowdonia have also failed to elicit any notable negative responses.

A study of visitors to the Horseshoe Pass (LUC 1990), an area of open moorland with substantial slate waste tips, found that the statements "The quarries and slate tips are an eyesore" and "All evidence of the quarries and tips should be removed" were disagreed with by a small majority of visitors who had not ventured as far as the quarries, and disagreed with by a 4 to 1 majority of those who had explored them. Only 3% of visitors listed the quarries and tips among their 'dislikes' of the area.

The public interest in industrial history is growing, and forms a growth area for tourism. This was recognised by the designation of 1993 as Industrial Heritage Year by the English Tourist Board (National Trust 1993).

It seems clear that there will always be differences of opinion between those who treat attractive landscapes as 'picture postcards' to be 'preserved' as natural views, and those who appreciate that Britain's upland landscapes are the product of man's influence and livelihoods. However, it would be incorrect to assume that modern practices in quarrying will maintain the current rural landscape any more than modern agricultural or timber production methods will do so. While these rural industries used the methods which created the landscapes we value, continuity was assured. The introduction of modern technology and financial constraints has provided the capacity for rapid change away from the familiar and towards an unpredictable landscape. It is that unknown and rapid change which most concerns the public, and this is recognised by bodies such as the National Trust, the Countryside Commission and the Countryside Council for Wales who promote traditional methods of land management and agriculture.

2.14   Hazards of disused workings

2.14.1   Hazards

Disused slate workings share many of the potential hazards shown by other rock quarries. These hazards, listed in Table 2.8, are generally self-evident and can be avoided by exercising due caution when exploring. The risks to the foolhardy and young children should not, however, be ignored, although hazards such as deep water and unprotected cliffs also exist widely in natural landscapes.

Slate workings have characteristics which distinguish them from most other quarries. The surface of slate waste heaps is composed of relatively large and unstable slabs which are easily disturbed, and when wet, slate is very slippery. Scrambling over waste heaps is therefore particularly hazardous. Although slate waste tips are inherently stable (section 2.7) there is the danger that the poorly-planned removal of material for use as fill could cause a collapse. Many abandoned workings also possess numerous structures in a more or less decayed state, which present hazards such as falling slates or more serious collapse.

2.14.2   Liability

The legal liability of occupiers or owners of such sites for any harm which comes to legitimate visitors or trespassers is set out in the Occupiers' Liability Act 1957 (Harte 1985). The duty of care owed to those specifically invited onto the land, or to those legitimately using public rights of way, is greater than that owed to trespassers.

Table 2.8 *Summary of hazards*

Quarries
- unprotected drops
- unstable or overhanging edges
- unstable faces
- deep water
- submerged rock edges
- submerged waste eg cars, fridges

Tips
- unstable blocks
- slippery surfaces
- unstable tramway tipping heads

Mine Workings
- roof collapses
- deep water
- hidden shafts

Structures
- unstable retaining walls
- unstable buildings, wheelpits
- unstable viaducts, bridges, inclines etc.

3   Slate Waste as a Resource

3.1   Uses of slate waste

3.1.1   Current uses

Slate waste has long been regarded as a resource without a market, and numerous attempts have been made to develop new uses. At present, the waste is processed and sold in a number of ways from sites of active production and by extraction from previously abandoned tips. The variable nature of slate means that not all slates are suitable for the uses noted here.

Slate powders

Slate powders are produced for use as an inert filler in many applications from face-powder to bitumastic coatings for undersea pipelines, and for the production of reconstituted slate tiles. Artificial slate roofing accounts for some 9 million m^2/yr; natural slate 2.5 million m^2/yr. (Harries-Rees, 1991).

Slate granules

Slate granules are produced for coating roofing felt. A range of granule colours is manufactured by Redland Aggregates at Blaenau Ffestiniog, using blue-grey and green slates and various pigments. Some green slate is quarried at Penrhyn Quarry specifically for granule production. The weight of granules and powders produced annually in Britain (about 40 000 tonnes) approximately equals that of roofing slate produced. The current demand of the UK granule and powders market is broadly satisfied by current production, but a modest increase in demand and therefore waste usage is forecast (Table 2.1).

Construction materials

Slate waste is crushed and graded for sale as construction material, and can be screened to produce any size range demanded. Typical applications are drainage blankets, pipe trench backfill, french drain fill, capping or blinding layers, and 'rip-rap' stone of 150mm upwards. Type 1 and type 2 sub-base materials meeting the Department of Transport (DOT) Specification are the largest uses of waste slate, consuming about 200,000 tonnes per year, on average. Larger waste slate blocks of up to 10 tonnes weight are sometimes sold as armourstone for river and coastal protection. Producers will therefore set these blocks aside for possible future demand.

Bulk fill material

Slate waste is a source of bulk fill for the construction of road embankments, industrial areas and similar landfilling works. Materials for these uses command a very low price and so transport costs restrict their use to the vicinity of production. Road schemes can utilise large quantities of fill, although designers aim to balance cut and fill wherever possible. In the early 1980s, Kirkby in Furness Quarry supplied 600,000 tonnes of waste as fill for the Greenodd by-pass. Penrhyn Quarry, Bethesda has supplied as much as 500,000 tonnes per year to the A55 North Wales coast road for various construction uses, although that project is now largely completed. Many reclamation schemes in Wales have used slate waste within the site for constructional applications. At Braichgoch, Corris, the 1976-78 reclamation scheme created a new embankment for the A487 trunk road out of the slate waste tips.

Traditional walling

In areas of Cornwall and Devon, and parts of Snowdonia, slate waste was traditionally used for field walls and earth-stone banks. There is some demand for waste slate for the renewal or replacement of these landscape features, particularly when road improvements are carried out. Tips may then be scavenged to provide suitable building material.

3.1.2   Other potential uses

Slate waste can be used for the production of expanded lightweight aggregates, low-grade concrete, mineral wool, and other materials but none of these applications has proved economically viable.

Slate quarrying also generates other waste rock and overburden which can sometimes be sold as construction stone. At Penrhyn quarry, the Fron Lwyd grit which overlies the slate has been sold as roadstone. At Kirkby-in-Furness, the shale/weathered slate overburden has been successfully made into bricks and could be used in this way if brick demand grew.

3.1.3   Scale of use

Despite these specialised, and locally large-scale, uses of slate waste the total consumption remains at a fraction of current waste arisings and is insignificant in relation to the quantities present in slate waste tips. Arup (1991) reported that slate waste stockpiles contained 400-500 million tonnes and annual arisings were 6 million tonnes. The annual utilisation of slate waste was 0.5 million tonnes.

3.2 Production and quantities of slate waste

3.2.1 Geographical distribution

The amounts of slate waste produced each year, and standing in waste tips, have been variously estimated in reports by Gutt et al (1974), the "Verney Report" (DOE 1976) and Arup (1991). All relied on estimates provided by site visits or quarry operators since no comprehensive survey data exist. However, the Arup data are likely to be sufficiently accurate for general comments to be made, and at any site the quantity of waste available for use is constrained by many factors which are likely to outweigh any inaccuracy in the data quoted.

Scotland

Estimated quantity: 50 million tonnes (Arup 1991).
This figure excludes waste known to have been tipped into the sea at Easdale and Ellenabeich in the 'slate islands'. Very little waste remains on land at these sites. Approximately 9 million tonnes is located at Ballachulish and has been reclaimed (McGowan 1982). The largest remaining site is at Aberfoyle, where an estimated 5-10 million tonnes of waste exist. Slate production at other locations did not reach such a scale, but taking the many smaller quarries together the total estimate of 50 million tonnes is considered reasonable. Slate production on a commercial scale has ceased.

Cumbria

Estimated quantity: 25-30 million tonnes (unverified).
The largest single deposit is at Kirkby in Furness where 20 million tonnes of slate waste exist (Brownlee, pc). Other significant deposits are at Tilberthwaite, Broughton Moor, Elterwater, Honister Pass and Petts Quarry near Ambleside, whilst the balance is made up of over 50 smaller quarries and trial workings. In 1993, significant tipping continued at Kirkby in Furness, Elterwater and Petts Quarry. A total waste production of 0.4 mtpa was estimated (Arup 1991).

Cornwall and Devon

Estimated quantity: 20 million tonnes (unverified)
The Delabole quarry has a waste deposit of about 15-16 million tonnes (Hamilton, pc), composed of slate waste and overburden mixed in roughly equal proportion. This is 'sterilised' from large-scale use by the production facilities sited on the old tip. Waste tips at other sites also consist of overburden and weathered slates of doubtful value for aggregate use. Waste is currently produced at Delabole (50,000 tpa) and on a smaller scale at a number of quarries in north Cornwall and Tavistock, Devon.

Clwyd

Estimated quantity: 3-5 million tonnes (unverified).
The largest deposits are at two quarries on the Horseshoe Pass, where an estimated 1-2 million tonnes exist. Smaller deposits remain at Glyn Ceiriog, Glyndyfrdwy, and approximately 50 smaller sites in the region between Corwen and Llangollen (Richards 1991). There is no current slate or waste production, but some small scale removal of fill material has taken place and a further permission was granted in 1994.

Gwynedd

Estimated quantity: 350-400 million tonnes (Arup 1991)
The vast majority of the slate waste within Britain lies in Gwynedd, and estimates of quantity are very approximate. Among the most significant sites or areas are:

Penrhyn Quarry, Bethesda 200-250 million tonnes;
Dorothea Quarry, Nantlle 10 million tonnes (Anon 1993);
Penyrorsedd Quarry, Nantlle 5 million tonnes (Roberts W. pc);
Remainder of Nantlle Valley 5 million tonnes (estimate);
Llanberis valley 20-30 million tonnes (estimate);
Gloddfa Ganol, Blaenau Ffestiniog 60 million tonnes (Roberts W. pc);
Cwt y Bugail, Blaenau Ffestiniog 20 million tonnes (Roberts W. pc);
Llechwedd Quarries, Blaenau Ffestiniog 125 million tonnes (Gwynedd County Council pc)
Corris area over 10 million tonnes (Arup 1991).

Several hundred abandoned sites are listed by Richards (1991). These range from unproductive trials to quarries at which over 1 million tonnes of waste remains, but to undertake a topographical survey and quantify this waste would be impractical.

The principal sites of waste production in Gwynedd are:

Penrhyn Quarry, Bethesda (Hughes B. pc)	3 million tonnes per annum
Blaenau Ffestiniog (Arup 1991)	0.75 million tonnes per annum

Dyfed and Powys

Estimated quantity: 10 million tonnes (unverified)
No separate data exist for sites in these counties, but the largest waste deposits are centred around the Pembrokeshire quarries at Glogue (SN 22 32), Gilfach (SN 13 27) and Rosebush (SN 07 29), the Llwyngwern quarry in north-west Powys, and at Llangynog in north east Powys (County Planning Officers, Dyfed and Powys. pc). There is no significant new waste production.

3.3 The feasibility of further use of slate waste

3.3.1 Construction uses

The use of slate waste as construction aggregate and bulk fill is currently by far the largest market for the material, and was considered the most promising area of growth in consumption (Arup, 1991). The Government has provided considerable encouragement for widening the scope of use, including sponsoring studies into:

- waste availability;
- the use of secondary materials and lower grade aggregates;
- economic and physical constraints on utilisation;
- the role of specifications;
- dual tendering.

Research sponsored by the Department of the Environment (BRE 1993) has reviewed standards and specifications, and has concluded that there is a lack of suitable specifications for some materials, particularly lower grade/secondary ones, which means they are precluded from use in some cases. Where they are accommodated, this is often not appreciated by potential users. Further work by (BRE 1994) is determining the extent to which a range of waste materials, including slate wastes, are currently permitted in standards and specifications, and is identifying what further action is needed to give these wastes the best opportunity of being used. The adoption of the EC Construction Products Directive (Anon 1988) and Public Works Directive (Anon 1993b) will require that any material which meets the relevant technical standards shall be acceptable for use in contracts procured by the public sector. Some slate waste products do meet DoT Type 1 and Type 2 sub-base specifications.

Slate intended for use as a construction aggregate must meet the specifier's requirements for crushing strength, frost resistance and other parameters. The proportion of fine particles generated by crushing the waste is greater with softer slate, and this generally reduces the suitability of the material. Additional screening and preparation can produce a satisfactory aggregate in some cases, but at an increased cost. Other slates are not suitable as a raw material for aggregate production because they splinter or flake when crushed.

For many years it has been the practice for minerals planners to encourage the use of mineral working tips or deposits as an alternative to borrow pits when new developments are discussed with tenderers, but present systems do not favour this. In the tendering system for new highways, for example, the tenderer has to find his own bulk fill and for price reasons a search is always made as near the project as possible. If the procuring authority were to specify a secondary mineral source, all tenderers would be working to the same cost basis and secondary sources would be used. It is understood that such specification could, however, be contrary to the EC Public Works Directive. Dual tendering systems, which involve pricing on both a borrowpit and secondary materials basis, have not worked due to the need to meet high specifications, keep the costs down and a lack of confidence in secondary materials.

3.3.2 Aggregate demands

The total consumption of primary aggregates in England and Wales reached a peak of 270 million tonnes in 1989. Demand declined to 212 million tonnes in 1991 but it is estimated that annual demand will rise to between 370 and 440 million tonnes by 2011 (DOE 1994). Ten Regional Aggregates Working Parties, composed of local authority Mineral Planning Officers, mineral operators and central Government representatives, seek to coordinate the resolution of regional imbalances in supplies and demand. Their reports assist the Government to prepare guidelines for aggregates provision in England and Wales which are published as MPG 6 (DOE 1989, 1994b, Welsh Office in preparation). All RAWPs have stressed the need to increase the use of secondary aggregates including slate waste. Of particular concern has been the need to transport large volumes of aggregate to the south-east of England, and the consequent resource and environmental pressures on "exporting" regions. RAWP advice to DOE suggests that the total aggregate demand projections from 1992-2006 cannot be completely met from primary aggregate supplies, and so the government has set out a policy of encouraging the greater use of secondary materials in construction. MPG 6(DOE 1994) states that,

"... in keeping with the Government's commitment to a sustainable approach to the supply of aggregates it is in the national interest that aggregates, and products manufactured from aggregates, should be recycled wherever possible. It is also important that where they are technically, economically and environmentally acceptable as substitutes for primary materials, mineral and construction wastes should be used. This can afford considerable savings of raw materials and can reduce the areas worked for new materials as well as those used for the dumping of wastes. Government policy therefore is to encourage the use of secondary and recycled materials in construction and it is committed to increasing significantly the level of use."

The Guidelines for Aggregates Provision in England and Wales, contained in MPG 6, are summarised in Table 3.1.

Demand for construction aggregates is currently met predominantly by primary aggregates, ie those materials extracted from reserves specifically for the purpose. There are five main sources of construction aggregates in Britain. In order of 1989 output, these are:

- land-won sand and gravel, 109m tonnes;
- crushed limestone, 106m tonnes;
- crushed igneous rock, 42m tonnes;

Table 3.1 *Guidelines for aggregates provision 1992-2006*

Region	Slate waste	Primary aggregate provision 1992-2006[1]	Secondary aggregate provision 1992-2006[2]	Demand 1992-2006[3]
South East	No	450 mt	140 mt	1270 mt
East Anglia	No	145 mt	15 mt	225 mt
East Midlands	No	715 mt	70 mt	540 mt
West Midlands	No	330 mt	55 mt	490 mt
South West	Yes	715 mt	60 mt	610 mt
North West	No	175 mt	90 mt	440 mt
Yorkshire and Humberside	No	340 mt	65 mt	430 mt
Northern	Yes	245 mt	35 mt	275 mt
ENGLAND TOTAL		**3115 mt**	**530 mt**	**4280 mt**[4]

Notes:

1. Provision for production of land-won primary aggregates in each Region, for consumption within and outside Region.

2. Provision for consumption of secondary and recycled aggregates in each Region.

3. Projected consumption within the Region of aggregates from all sources.

4. Total includes marine-dredged sand and gravel 315 mt, imports from Wales 160 mt and imports from elsewhere 160 mt.

    Regions for which provision exceeds demand are anticipated to export aggregates to other regions.

    Guidelines for aggregates provision in Wales were being prepared by the Welsh Office at the time of this report (1994).

DATA FROM   GUIDELINES FOR AGGREGATES PROVISION IN ENGLAND MPG 6 (DOE 1994B)

- crushed sandstone/gritstone, 22m tonnes;
- marine dredged material, 21m tonnes.

During the 1960s, 1970s and 1980s there was a sustained increase in the consumption of crushed rock whilst output of sand and gravel remained relatively static at around 100 mtpa. In 1989/90 the estimated use of waste materials for aggregates purposes was 10% of the total for primary aggregates, *ie* 32 million tonnes.

The South East is the largest regional market for aggregates. Despite having relatively extensive resources of land-based sand and gravel, local supplies are sufficient to meet less than half of market demand. This is because it is becoming increasingly difficult to develop available deposits for various reasons, including environmental constraints. The regional imbalance of supply is made up by imports from rock quarries, mostly in the East Midlands and South West regions, and from marine sources.

3.3.3   Transport of aggregates

Aggregates are heavy and demanded in bulk, and so haulage costs may account for 50% of the final sale price. Road transport is the prime method of hauling aggregates owing to its flexibility and graduated cost structure. Road transport is also used for onward distribution from bulk rail or marine depots to the final customer. Typically, transport costs restrict the supply of aggregates by road to 20-30 miles. This favours smaller, local producers in areas where demand is insufficient to warrant rail haulage. Rail haulage is used to move large quantities of crushed rock over long distances. At present, about 20% of all aggregates are moved by rail. Generally the rail transport of crushed rock seems set to expand, supported by improved wagon technology and increasing demand for higher value, premium materials. Sea transport is used particularly by those coastal quarries which can load directly into ships, and for long distance transport to bulk depots in the south-east.

Slate quarries, both active and disused, are all located in areas of relatively low population density and low constructional activity - the Scottish Highlands, the Lake District, north and mid Wales, Pembrokeshire and north Cornwall. The primary areas of demand for aggregates and fill are the densely populated areas of the south-east, the midlands, the north-west and central lowlands of Scotland. At present, therefore, the demand for slate waste is low and other primary aggregate sources are better placed to meet aggregate demands, except in the immediate vicinity of slate workings.

3.3.4 Scenarios for increased use of slate waste

It has been suggested (Arup, 1991) that economic mechanisms might be used to alter the balance to favour the increased large-scale use of slate waste and other secondary aggregates. This would tend to increase slate waste utilisation in two distinct categories:

1 'Local' use. The use of waste in low-value applications such as bulk fill, within the area over which a price advantage existed. As road haulage forms a large part of the price of such materials, the area would be dictated largely by the relative location of other sources of supply. Operations could be intermittent to supply larger local projects. Waste could also be used, after processing, for higher value applications such as graded aggregates provided that sufficient demand existed to sustain the investment required.

2. 'National' use. The processing of waste into higher-value specified products such as DOT Type 1 and Type 2 sub-base, for which long-distance transport to areas of higher demand would be economic. This would require investment in processing and transport facilities to achieve economy of scale, and would therefore have to be a permanent continuous operation. Existing, active slate quarries would be better placed to meet this demand.

The 'local use' scenario could occur at almost any slate waste deposit, subject to environmental and economic constraints. The 'national use' scenario would be restricted to larger waste deposits with access to economic bulk transport facilities, ie rail or sea. Figure 3.1 shows the location of the larger slate waste deposits identified in section 3.2 in relation to active rail lines and port facilities. It can be seen that Kirkby in Furness, Penrhyn, and Blaenau Ffestiniog are the best placed to use these facilities and are therefore the most likely sites of any large scale increase in waste utilisation for 'national' consumption.

Penrhyn Quarry, Bethesda, was the subject of a study reported to the DOE (Arup 1991). The owners, McAlpines, have evaluated transport options and carried out trial 'exports' of slate aggregate to the south-east of England using road haulage between the quarry and existing port facilities at Bangor. The case study evaluated two options for full-scale exports via purpose-built loading jetties, and concluded that these were practical and would become economically viable if modest grants towards the capital investment were available.

Permission for the large scale removal of slate waste from Dorothea quarry, Gwynedd, was refused in January 1993 on historical, environmental, traffic and amenity grounds (Box 3.1).

Smaller-scale proposals for slate waste removal for 'local' use have been permitted in the Nantlle valley and elsewhere. These operations rely entirely on road transport, and the specific consequences of this have been scrutinised carefully in each case, since most slate workings are reached via relatively minor roads.

3.4 Development control over waste extraction

3.4.1 Waste extraction as 'development'

The removal of slate waste from a slate working site or tip may be allowed:
a) when planning permission exists specifically, or
b) under the Town and Country Planning GDO 1988. There are, in Part 23 of Schedule 2, three situations or classes where material can be taken from an existing deposit of mineral waste:

1. Class A would rarely apply today as it provides exemption to sites or areas of tip where mineral was being removed during a 12 month period prior to Section 1 of the 1981 Act coming into force and for a 6 month period thereafter ie until 19 November 1986. Approval continues, however, if at that time a planning application was made and has yet to be determined by the planning authority.

2. Class B allows for the removal of material from a stockpile.

3. Class C allows for the removal of material from small (less than 2ha) or temporary mineral working deposits (where no material has been on the land for a period exceeding 5 years) subject to 28 days notice being given to the planning authority. The details to be given in this notice are amplified in MPG 5 (1988d). Under certain circumstances the mineral planning authority can direct that an application for planning permission be made.

All other cases of removal of material from a tip or mineral working deposit require planning permission, as does the removal of material from disused mineral railway embankments and similar engineered structures.

Figure 3.1 Slate waste and active rail links, 1994

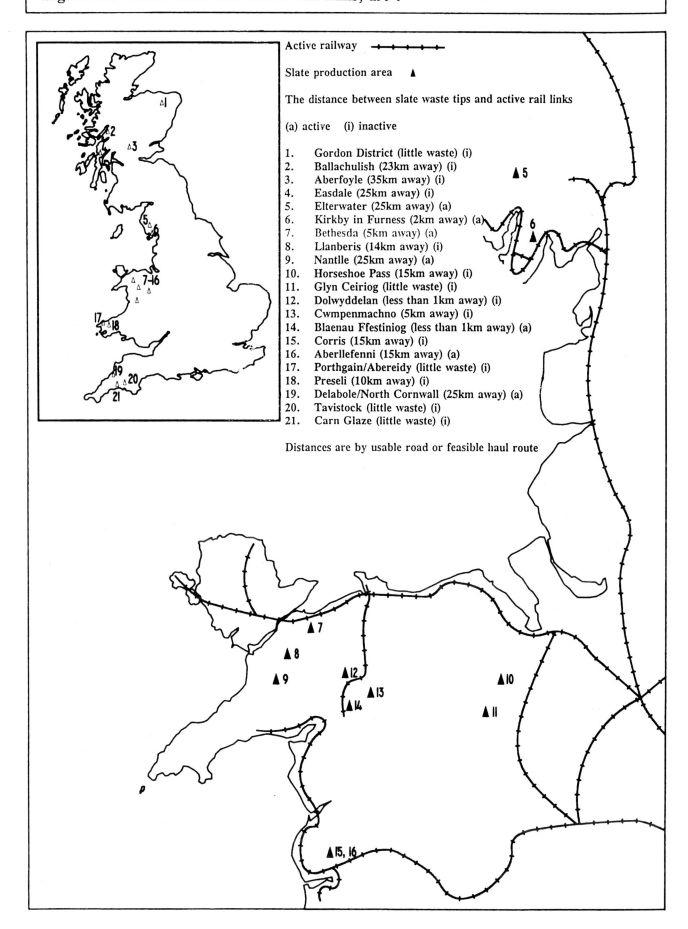

Active railway ———+———+———+———+———

Slate production area ▲

The distance between slate waste tips and active rail links

(a) active   (i) inactive

1. Gordon District (little waste) (i)
2. Ballachulish (23km away) (i)
3. Aberfoyle (35km away) (i)
4. Easdale (25km away) (i)
5. Elterwater (25km away) (a)
6. Kirkby in Furness (2km away) (a)
7. Bethesda (5km away) (a)
8. Llanberis (14km away) (i)
9. Nantlle (25km away) (a)
10. Horseshoe Pass (15km away) (i)
11. Glyn Ceiriog (little waste) (i)
12. Dolwyddelan (less than 1km away) (i)
13. Cwmpenmachno (5km away) (i)
14. Blaenau Ffestiniog (less than 1km away) (a)
15. Corris (15km away) (i)
16. Aberllefenni (15km away) (a)
17. Porthgain/Abereidy (little waste) (i)
18. Preseli (10km away) (i)
19. Delabole/North Cornwall (25km away) (a)
20. Tavistock (little waste) (i)
21. Carn Glaze (little waste) (i)

Distances are by usable road or feasible haul route

Box 3.1 *Dorothea Quarry - a major secondary aggregates proposal*

The proposal

The Dorothea quarry complex contains some 10,000,000 m³ of slate waste in many waste tips over a total area of 120 ha. Quarrying ceased in 1970 although the planning permission granted in 1951 is valid until 2042. The applicant sought permission from Gwynedd County Council, the Mineral Planning Authority, to extract and process up to 9.5mt of slate waste over 20 years. This would generate up to 8 mt of aggregate and 1.5 mt of fines. The aggregate would be transported by road to Penrhyn harbour, Bangor (200,000 tpa) and Porthmadog harbour (100,000 tpa) for shipment to south-east England and continental Europe. The remaining waste would be recontoured to create parkland and woodland using the slate fines and re-exposed soils as the growing medium. The importation of domestic refuse for disposal in the quarrying holes had been discussed locally, but did not form part of the planning application.

Planning issues

The Mineral Planning Authority considered the issues raised by this application. They included:

- the persistent demand for aggregates in the south east of England, and the role of Gwynedd as a supplier;
- the demand for secondary aggregates derived from slate waste, which was identified by the Arup study (Arup Economics and Planning, 1991b) as a potential source;
- the availability of alternative aggregate sources in Gwynedd. The permitted reserves of primary aggregates are 512 mt (1991 estimate) and there are some 300 mt of slate waste in the county. The varying characteristics, particularly accessibility, degree of revegetation and visual obtrusiveness, make the slate deposits difficult to compare with each other;
- the environmental implications of slate waste removal on the Dorothea Quarry site, and of material processing and transport on the wider area.

The environmental implications

The proposal would have generated substantial short and long-term effects:

- 110 HGV movements per day on the B4418 through Penygroes and on the A487 north to Bangor (24km) or south to Porthmadog (20km);
- disturbance to residents in 29 houses lying within 100m of the tips to be removed;
- disturbance of the established wildlife of the site;
- removal or regrading of over two-thirds of the slate waste within the 120 ha site, completely altering the existing landform;
- removal of much of the setting of two Scheduled Ancient Monuments and many industrial archaeological features in the centre of the site;
- removal of much of the topographical interest of the site, reducing its potential as a 'quarry park' as proposed by the Borough Council.

The County Council has acknowledged the need to encourage the use of secondary aggregates where appropriate, but considered the availability of alternative reserves, the environmental implications, the concerns of statutory and non-statutory consultees and residents, and the long-term future of the site. The effects on features of archaeological and historic interest, on nearby residents, traffic generation and on the potential for the conservation of a representative selection of industrial archaeological sites, among other planning considerations, led the County Planning Officer to recommend that permission be refused. The Secretary of State for Wales issued an Article 14 Direction not to grant planning permission and in January 1993 permission was refused.

3.4.2   Planning permissions for waste removal

The process of application for permission to remove any mineral waste is similar to that for a new development. It is likely that the planning authority will pay particular attention to transportation issues including site access, and to any features of ecological or historical interest on the site. Any permission granted will also contain conditions regarding the limitation of nuisance, time restrictions, restoration and other relevant matters.

Section 106 agreements, termed "planning obligations", are increasingly adopted to cover matters, such as traffic routing, highway cleansing and guarantees relating to the long-term management of the restored site, which fall outside the scope of planning conditions. The variable quality of slate waste tips can cause unexpected problems for operators, and potentially render a project unviable. The relative uncertainty and low value associated with mineral reworking increase the risk of non-compliance with planning conditions.

3.4.3   Development and local planning

In all cases where some new use or development is being considered, the developer is advised to contact the local planning authority or mineral planning authority at an early stage. Informal discussions can help to identify the sites or uses which are least disturbing to areas of value and most compatible with planning objectives or reclamation strategies for the area.

NOTE Where views on the interpretation of the law are expressed these should only be taken as a guide; for more definitive interpretation the original documentation should be consulted.

3.5   Environmental consequences of primary and secondary mineral working

3.5.1   Comparison with primary minerals

The extraction of slate waste from tips, its preparation for sale, and its transport from source to site of use, are similar in many ways to the extraction, preparation and transport of primary materials. The environmental effects of surface mineral workings were studied in recent research (Roy Waller Associates 1991) which reported that the main concerns with large stone quarries were visual, traffic, dust, noise and vibration (including blasting). Surface water contamination and industrial archaeology are current concerns for slate waste extraction. The loss of land taken for extraction or tipping was also perceived as an undesirable but inevitable environmental consequence of mineral working. That research produced published guidance describing the concerns, identifying good practice and suggesting further topics for research. The general concerns and good practice are not therefore repeated here.

In this section, the use of slate waste in principle is compared with primary mineral extraction under the main environmental headings, and then an environmental summary is given for each of the larger areas of slate working.

3.5.2   Visual impacts

Visual impacts include:
- loss of existing landscape features;
- intrusion of alien features from quarry activity;
- obstruction of views to and from unaffected areas.

Primary mineral extraction can cause all of these impacts to an extent which is dependent on location, working method and ancillary activities, mitigation measures and restoration. Quarrying activity currently takes place in areas designated as National Parks, Areas of Outstanding Natural Beauty or other special landscape status, and visual impacts in particular form a major item of conflict during the planning process. The use of slate waste as an alternative material would reduce the demand for primary materials and would therefore reduce the extent of the associated visual problems. However, the main areas of slate working are all within, or adjacent to National Parks, Areas of Outstanding Natural Beauty, Heritage Coast or Forest Park (see Figure 2.3) and so equal scrutiny would be applied to reworking schemes. In principle, short term operations to remove tips and restore the land to desirable new landscapes might generally be regarded as visually beneficial, but it is known that the public frequently regards even short-term disturbance as more unwelcome than the familiar sight of slate waste tips (Williams, 1992). The use of slate waste from active quarries such as Penrhyn or Delabole has no visual effect since at Penrhyn waste is removed from massive tips in the centre of the complex and at Delabole current production wastes are used. At Penrhyn production waste could provide up to 3mtpa of secondary materials, with no effect on existing tips.

3.5.3   Traffic

There is no significant difference between the transportation of slate waste and any other rock or aggregate mineral, except in the location of origin and routing of traffic. In section 3.3 the practical constraints of slate waste transport were discussed, for 'local' and 'national' use scenarios.

Vehicles carrying minerals on roads close to mineral workings are among the heaviest and possibly the largest to use such roads. The vehicles are often out of scale with the rural and urban roads that they have to use, especially in the vicinity of the workings and the customer's site. For example the access to one major site is along a narrow road through a residential area with limited visibility and passing places. Even low traffic flows in sensitive areas give rise to complaints.

Empty lorries may cause more disturbance than fully loaded ones; they tend to travel faster and to be noisier because they suffer from body-slap when going over bumps, or speed-control humps. Turbulence in the empty bodies of unsheeted vehicles may scour out dust (Roy Waller Associates 1991).

3.5.4   Dust

The extraction and crushing of any rock mineral can potentially cause dust. The raw material in the tips retains some dampness in dry weather, but slate crushing plant such as that used at Penrhyn Quarry, Bethesda, is operated with the addition of water to the raw material feed, to prevent dust emission. The Environmental Protection Act will introduce further control procedures. Dust can also be generated by vehicles using site roads. Slate does break down to dust in such circumstances and so routine water application is required. At Penrhyn Quarry, haul roads are being surfaced with road planings when available, to reduce dust generation. Even in the relatively wet western areas of Britain in which slate workings are found, dust-raising, ie windy, dry days form 25% of winter days and a greater proportion of summer days. Under these conditions, the application of good practice can eliminate the transport of wind-borne dust off-site (Hughes T. pc).

3.5.5   Noise

The extraction and preparation of slate waste and primary minerals involve essentially the same processes, except that slate waste extraction requires no blasting. Some slate wastes from older tips have a relatively small particle size and so require less crushing than the equivalent hard rock material. Slate requires lower crusher impact pressures than hard rocks (Arup, 1991). The loading and transport machinery is common to both mineral sources. Noise generation and propagation therefore is a site specific matter rather than a mineral-specific matter. Noise impinges on residents, recreational users of nearby land, and wildlife. The significance of a given noise intensity pattern depends on the location and numbers of people affected, and the wildlife of the locality. Guidelines given in MPG11 (DOE/WO, 1993) provide advice on how planning controls and good environmental practice can be used to keep noise emissions to environmentally acceptable levels.

3.5.6   Vibration and blasting

Blasting causes ground vibration, air overpressure waves and audible noise, which can be physically or psychologically disturbing to those exposed. The extraction of slate waste does not involve blasting, although the extraction of 'virgin' slate by blasting continues at some potential sources of slate wastes. If primary mineral extraction was reduced by the use of secondary minerals, then some reduction in blasting at primary sites might be expected.

3.5.7   Surface water pollution

Slate waste is chemically inert unless pyrite is present, and pyritic slate is unlikely to be acceptable for most purposes. The only significant effect of slate waste extraction on surface watercourses would be from particulate deposition, but this can successfully be controlled by the interception of surface water, wheel-wash water and other 'dirty' water for treatment by filtration or settlement. Similar techniques are required at most modern mineral extraction sites.

3.5.8   Loss of land

At first sight the major difference between the extraction of primary minerals and the use of slate waste is that the former takes undisturbed land, some of which is of nationally recognised wildlife importance, whilst the latter uses derelict, tipped land or uses current production wastes and thereby reduces the use of undisturbed land for new tipping. There are many cases where this holds true, but two factors should be considered:

1. Some of the land affected by slate waste tipping now has a landscape, wildlife, recreation and historical value of its own, as described in sections 2.9-2.11.

2. Without the continued tipping of production waste as quarry backfill at larger sites such as Delabole and Penrhyn, the eventual problem of quarry restoration may be increased by the lack of stable fill material.

The south-east region of England is seen as the main 'national' market for increased supplies of secondary minerals. The primary minerals which currently supply that market are limestones extracted from the Mendips of Somerset and igneous rocks extracted from Leicestershire (Arup 1991). Sand and gravel extraction within the south-east is already constrained by the difficulties in meeting current environmental expectations, but the Somerset Mendips are also a recognised high quality landscape and are an important aquifer (Moon 1992). On a more local scale, a limestone geology often produces a species-rich flora of nature conservation interest, which would be severely damaged or lost by quarrying activity.

There are few, if any, sites of potential primary or secondary mineral extraction at which the value of the land is negligible.

3.5.9   Loss of industrial archaeology

It is possible that primary mineral extraction will damage or destroy archaeological remains and the remains of earlier mining or quarrying, although primary mineral quarries tend to be in unworked sites and archaeological remains are offered some protection by guidance in PPG

16: Archaeology and Planning. The removal of slate waste from abandoned or historic slate workings will almost inevitably lead to the loss of some features of interest and will reduce the setting and understanding of features which remain. This loss is a significant disbenefit of some proposed slate waste extraction projects, and was one reason for the refusal of planning permission for the Dorothea proposal (See Box 3.1).

3.5.10 Summary

For the 'local' use scenario set out in 3.3.4, the site specific characteristics are likely to outweigh any general comparisons between primary sources and slate waste. For the 'national' use of slate waste there are significant advantages in terms of reduced land 'take' and possibly, reduced blasting disturbance. The local effects of transport mode and visual disturbance should be considered site by site, although a substantial increase in the use of current production wastes could be achieved with no visual disbenefit. Methods for the assessment of individual sites are described in section 4.

3.6 Opportunities to reshape or reclaim tips and quarries

3.6.1 Mineral extraction and reclamation

A carefully designed scheme of slate waste removal for use as a secondary mineral can also act as a scheme of reclamation or rehabilitation for an abandoned working, particularly if the waste removal is governed by planning permission conditions. In principle, the sale of some material during a publicly funded reclamation scheme could offset the cost of works, just as coal recovery can offset the cost of reclamation works to colliery spoil heaps. Where mineral sale was the primary objective of a privately sponsored scheme, reclamation could be achieved at no public cost through planning conditions or a legal agreement. In practice, slate waste has a very low mineral value and so the potential scale of revenue-funded works is substantially lower than that from coal recovery schemes.

The complete removal of an area of slate waste tips offers scope for restoration of the land to new land uses. For example, the planning application for waste removal at Ty Mawr East, Nantlle (Richards, Moorehead and Laing 1992) contained proposals to restore the 4 ha site to agriculture and woodland, at no cost to public funds. This application was approved.

The partial removal of slate waste tips can also be beneficial if the removal is designed to leave a landform which suits the eventual land use, or contributes to a reclamation scheme. As part of the reclamation of the derelict Glyn Rhonwy site, Llanberis, 45 000 mt of slate waste were extracted by a developer and crushed to prepare 125mm down and 75mm down aggregate for use in

Caernarfon as a construction fill. This operation left an additional development platform at Glyn Rhonwy at no cost to the reclamation scheme. The planning application for large-scale slate waste removal from the Dorothea complex, Nantlle, sought to extract about 60% of the 10 million tonnes of material estimated to be present, and proposed that some of the remainder would be used in site restoration (Anon 1993).

3.6.2 Additional benefits

Waste removal schemes can contribute to reclamation in other ways. In 1990, slate waste from an active Blaenau Ffestiniog quarry was transported to Penrhyndeudraeth to fill soft land in preparation for a publicly funded industrial development. The unsuitable peaty soil removed from the site was transported back to Blaenau Ffestiniog in the returning wagons for use in future slate waste revegetation work. It would otherwise have been disposed of.

3.6.3 Landfilling and reclamation

The Ty Mawr slate waste removal scheme (section 3.6.1) was linked to proposals for landfilling the quarries with inert builders' wastes. The applicants intend to segregate any subsoil or soil-forming materials from these wastes, for use in site restoration (Richards, Moorehead and Laing Ltd 1992). In this way the quarries as well as the tips would be restored to the original land profile and to new uses. The wider adoption of landfilling is, however, constrained by the localised availability of inert wastes requiring disposal and the environmental concerns associated with leachate or gas-generating wastes. Waste disposal is discussed further in section 4.2.

3.6.4 Integration of mineral extraction and reclamation programmes

The potential benefits of the integration of slate waste use and reclamation/rehabilitation programmes will be maximised by the adoption of a plan-led approach, in which opportunities for positive spin-offs and linkages are identified. In this way, operators seeking secondary mineral reserves can be guided towards schemes which have wider benefits, and the detailed operating plans can be refined to achieve suitable landforms, conserve soils and soil-forming materials, conserve desirable site features and vegetation, and minimise the subsequent costs of preparing the site for its new use. Section 6 describes a framework for the assessment of an area of slate workings, as part of the preparation of an integrated plan.

3.7 Protection of secondary mineral reserves from sterilisation

3.7.1 Future demand

In section 3.3 the Government policy of encouraging secondary aggregate use was reported. Although the majority of this demand may be met from active workings' waste production, which is expected to increase when primary slate consumption resumes its increasing trend, some local demands will be met from waste deposits at various locations. There is therefore a need to identify the most suitable deposits, ie those where least environmental damage could be caused, and to protect these from sterilisation by development or curtailed access.

3.7.2 Safeguarding of reserves

MPG1 paragraph 31 gives guidance on the manner by which potential reserves of mineral including slate can be safeguarded against other types of development. Minerals Planning Authorities are required to supply Local Planning Authorities with plans showing those areas in which they wish to be consulted when developments are proposed. Policies should also be included in Development Plans indicating areas where protection is seen to be paramount. A typical policy as seen in Development Plans for a slate producing area is, "Within or adjacent to areas believed to contain reserves of ... slate ... or to be necessary for their exploitation there will be a presumption against non-mineral development which would lead to the sterilisation of the reserves or which by virtue of its siting or nature would not be compatible with mineral workings or associated operations". It is also suggested

in MPG 1 that policies to allow the extraction of minerals before other development takes place may be desirable, for example, mineral resources which would be sterilised by road construction. There should also be a clear distinction between areas to be safeguarded for the future and those areas within which there will be a presumption in favour of mineral working within the lifetime of the Development Plan.

3.7.3 Protection of rail links, port facilities and access routes

Most Development Plans contain policies encouraging the use of rail, marine and internal waterway transport systems for minerals, and discouraging the use of congested or minor roads. The protection and preservation of existing facilities, and where appropriate the improvement of trunk roads, are therefore of considerable importance if large volumes of slate waste are to be transported or if the backfilling of disused quarries with imported materials is to be undertaken. Bethesda lost its rail link in 1963, and so all production from Penrhyn Quarry is now transported by road. The Delabole rail link closed in 1962, and now all materials are also transported by road through the village. Substantial increases in the production of slate aggregates would therefore have major road traffic impacts. British Rail's line to Blaenau Ffestiniog was, in 1994, widely regarded as under threat of closure, a step which would greatly reduce the potential of Blaenau Ffestiniog to supply large quantities of material without unacceptable road traffic. Existing rail-head and port infrastructure, or land for its construction, should also be identified and protected if necessary. Access routes which avoid residential areas could be protected from development.

4 Alternative Uses of Former Slate Workings

4.1 The case for conserving selected workings

4.1.1 Education

Examples of man's past activities serve a valuable role in educating current generations about the lives and industries of earlier generations. Many good examples of old activities were lost or destroyed at a time when they were regarded as damaging. Buildings with industrial functions have rarely been protected by listing since few have been perceived as possessing any aesthetic merit (AIA, 1991). Since it is impossible to tell with any certainty what future generations will regard as interesting or valuable, there is perhaps a philosophical argument that some examples of any current or recent activity should be conserved so that future generations may have the opportunity to study them. The White Paper 'This Common Inheritance' (Anon, 1990b) states that something of each major industrial advance ought to be retained as a record of achievement, as an inspiration for the future, and a lesson in the nature of economic progress.

Industrial archaeology, ecology, geology and other areas of study continue to develop new techniques, new theories and new knowledge. Studies in future years may well learn more, or revise existing understanding by the application of these new techniques, theories and knowledge. Selected sites should therefore be conserved to allow further study and re-evaluation in the decades and centuries to come. However, studies and records of sites are at present far too incomplete to allow a fully informed selection to be made. Until surveys and evaluations have been completed, many experts will take a protective approach. The Welsh Industrial Archaeology Panel was in the process of drawing up a preliminary list and evaluation of the most important slate sites, in 1994 (Williams M. pc).

4.1.2 Conservation of sites

Very few abandoned slate working sites currently receive any form of conservation or care. The likelihood is that many sites will continue to be left 'as they are', but in reality this is equal to allowing them to deteriorate. Over time such sites slowly change from being localities with recognisable structures and features to being places with amorphous piles of rubble. Not being of a great size and not being close to habitations these sites pose little threat and are likely to be left to disintegrate slowly and be 'reclaimed by nature'.

There are a great number of small sites in places such as the Snowdonia, Lake District and Pembrokeshire Coast National Parks and the North Cornwall Heritage Coast. The authorities responsible for these areas actively manage some of these sites, which are part of the physical manifestation of their cultural heritage. Section 4.9 gives examples of abandoned slate workings which are seen as positive elements in a fairly low-key interpretative strategy for the National Parks.

4.1.3 Continued supplies of special slates

Some workings, or particular sections of old workings, are considered valuable as the potential source of new slates of a particular colour or appearance for use in renovation work on historic buildings. Bodies such as English Heritage have expressed the view that the adoption of new uses for slate workings may prejudice the opportunity to reopen closed quarries on a small scale to provide slates for restoration projects (Fidler J. pc). Attempts have been made to recommence slate working on the Hill of Curskie, Grampian, to help satisfy the growing demand for Scottish slate in new and conserved buildings. This slate has a unique colour and texture which cannot be matched from other sources (Barker J. pc). In Cornwall, the county planning department have conducted a feasibility study into a co-operative marketing approach for the small slate producers, to assist in the maintenance of this industry and the supply of local materials (Jones M. pc). A parallel exists in the Pennine uplands, where stone slates were traditionally used for roofing but have not been quarried for about a century. As a result of concerns that no supplies for conservation work exist, Derbyshire County Council organised a seminar to explore and encourage the recommencement of a small quarrying industry to enable the buildings and landscapes of the Peak District to be maintained (Derbyshire County Council, 1992).

4.2 Waste disposal in slate workings

4.2.1 Waste disposal: inert fill

There is great interest in mineral excavations including slate quarries as possible disposal sites for waste.

In general, there should be no water quality or air quality concerns about the deposition of slate or other genuinely inert waste in slate quarries. By definition, inert fill does not affect the chemistry of water passing through it, and non-pyritic slate waste would fall within the inert category. Clay and silt-sized particles arising from slate or soils could give rise to increased turbidity and sediment in water courses, to the detriment of aquatic organisms. Fine-grained material such as that dredged from settlement lagoons, is likely to be found within recent slate waste tips. Some construction companies are permitted to dispose of demolition rubble and builders' waste by landfilling at quarries in North Wales. If such waste contains a large proportion of lime mortar, plaster or similar constituents it could alter the pH of drainage water in areas of naturally acidic strata. In addition, waste timber, paper and cloth will decay with the possible formation of landfill gas. Controls are therefore usually placed on the composition of the "inert" waste by the Waste Regulatory Authority. It is government policy to encourage the recycling of construction wastes rather than their disposal (DOE 1994b).

Dust problems could arise if fine-grained waste slate or other materials were not properly covered immediately after tipping.

4.2.2 Waste disposal: domestic refuse and special wastes.

Before the implementation of the Control of Pollution Act 1974, a number of slate workings were used for the deposition of domestic wastes. Many small workings were used as convenient disposal points for locally produced wastes and because of their restricted volumes they tended not to cause serious water pollution. At that time the implications of water pollution were not generally recognised and adequate legislation was not in force. Derelict slate quarries were considered attractive disposal points for many wastes that are currently classified as 'special waste' under the Control of Pollution (Special Waste) Regulations 1980. Twll Ballast in the Nantlle Valley received a range of industrial wastes in the period 1963 - 1974 (W S Atkins Environment 1992) and the nearby Cilgwyn quarry complex received unidentified wastes before being taken over by the local authority in 1974. Other quarries remote from settlements have attracted fly-tipping, particularly of old vehicles, furniture and garden wastes.

Bowithick quarry in north Cornwall was used for refuse disposal for several decades and continued, under local authority management, to receive local domestic and commercial wastes until it was completed and capped in 1992. Nearby slate waste tips were used as part of the capping material.

4.2.3 Suitability of slate quarries

The suitability of slate quarries for the disposal of leachate-producing wastes in accordance with modern standards is highly questionable. Domestic wastes can produce a leachate that is as polluting as that from many 'special' wastes. The slate areas of western Britain receive a relatively high rainfall and unless the site is sealed, leachate will be generated in large volumes. Whereas "dilute and disperse" was the accepted policy for such leachates until the late 1970s, the philosophy of waste disposal has now changed completely to the containment of leachates and minimisation of leachate generation.

Slate is essentially impermeable, but water movement occurs because of fissuring or fracturing. Field investigations to quantify such permeability are expensive and frequently inconclusive. The conventional concept of a water table is also generally invalid within the Palaeozoic rocks associated with slate workings, since groundwater usually exists and migrates within discrete fissures. Boreholes in these strata may or may not record water according to the fissures encountered. Many pit quarries were kept dry by pumping during their working lives, and filled with water after abandonment. Twll Ballast and Bowithick are known to have been partly flooded when tipping commenced. The permanent water level in a quarry may not reflect a 'water table', simply the balance between surface and groundwater inputs and losses through drainage and evaporation.

Some quarries are drained by drainage adits or former access adits and tunnels which can be impossible to trace. The NRA objected to the use of the Abercwmeiddau slate quarry, Meirionnydd for the disposal of leachate-producing wastes mainly because it was not known where the various tunnels in the bottom of the quarry led to, or how effectively such tunnels could be sealed (Thorpe M. pc). It can be extremely difficult to prove that a dry, apparently-enclosed quarry is not actually drained by unknown adits or by permeable strata resulting from rock fracturing. Attempts at water tracing may well prove inconclusive.

4.2.4 Site lining

Modern planning and site licensing conditions would require a waste disposal site to have its base and sides sealed. This is unlikely to be a viable option for most slate quarries because of the steepness and jagged nature of the sides, and the lack of suitable locally-occurring natural materials for lining of the quarry. Local clays tend not to occur in either adequate volumes or qualities for lining slate workings. Artificial membranes are likely to prove both unsuitable and too expensive. The only site that is known to be currently licensed to receive domestic wastes, consists of three interconnected former slate quarries. Two of the quarries are now nearly full and are unlined. As a result of negotiations in 1992 the NRA require that the third quarry be lined so that leachates can be collected and disposed of in an acceptable manner.

4.3 Industrial and other development

4.3.1 The suitability of slate workings

Some slate workings have been reclaimed to provide land for industrial, residential or commercial development. Other workings might provide suitable locations for industry and for "bad-neighbour" land uses if tips of slate waste or the quarry holes concealed the users from their surroundings. Activities such as car repair and vehicle dismantling are undertaken in some workings, where buildings have been adapted and noise and untidiness are isolated from the general public. Many such isolated workings are reached only by very poor roads, and so they are unsuitable for uses which would generate significant volumes of traffic. Utility services are often lacking, and the cost of installing connections to distant sources of supply would make the development of these sites very expensive.

4.3.2 Impacts on surroundings

Development on a substantial scale will have an impact on its surroundings. Where development is proposed for former slate workings the impact on desirable site features such as wildlife, industrial archaeology and other uses of the site should be considered in the same way as for a greenfield site. The development of derelict workings may cause less adverse impacts than development elsewhere, but this should be demonstrated rather than assumed.

4.4 Sports and leisure uses of slate workings

4.4.1 Legal liability

A variety of outdoor sports and leisure or recreational activities are conducted in abandoned slate workings where these are either well suited to the activity or the only available site. A list of activities and known examples of locations is given in Table 4.1. In the majority of cases the activities do not have the formal approval of the landowner and are at best tolerated. Landowners, and in particular public or commercial bodies, are very cautious about the legal liabilities which may be attached to any form of open or tacit approval of these activities, although one report (Land Use Consultants 1992) suggests that provided suitable notices warn of the potential dangers and make it clear that liability rests with the site users, access can be

Table 4.1 *Outdoor activities in abandoned slate workings*

Activity	Examples of location
Field nature study	Many
Geology: study and teaching	Geological Conservation Review sites in Wales. Fossiliferous slates in Gordon District, Grampian.
Mountain bike/off-road motorcycling	Moel y Faen, Clwyd
Adventure training/'team challenge'/outward bound groups	Aberfoyle, Stirling Cwm Machno quarry, Gwynedd
Rock climbing/abseiling	Abercwmeiddaw, Meirionnydd Allt Ddu, Vivian and others, Llanberis Hodge Close, Cumbria Birnam, Stirling
Caving/mine exploration/ underground industrial archaeology/rescue training	Moel y Faen, Clwyd Honister Pass, Cumbria Rhiw Bach quarry, Gwynedd
Sub-aqua (recreational and training)	Hodge Close, Cumbria Dorothea, Gwynedd
Small-bore rifle shooting	Alexandra, Gwynedd

Note: Many of these activities do not have the permission or approval of the site owners.

permitted except by way of business. If the public are invited to use the land eg in a country park, then liability would rest with the occupier.

4.4.2 The attraction of slate working sites

The general requirements of many outdoor recreational activities were set out in 'The Use of Land for Amenity Purposes' (LUC 1992) and so are not reproduced here. Abandoned slate workings offer a number of specific attractions to certain groups of recreational users. These include:

- dramatic topography and dry ground conditions for off-road motor-cycling and mountain bikes;
- quarry faces providing rock climbing at low level, for a range of abilities;
- deep sheltered water for winter sub-aqua diving when sea conditions are unsuitable;
- easy road access for users, avoiding long walks with equipment;
- containment of noise and stray shots for organised gun club activity.

4.3.3 The views of sports organisations

The governing bodies and clubs which organise such activities have expressed their concern that such facilities should remain available for use. Slate quarries in the Llanberis area are included in a climbing guidebook (Davies K. pc), and have been developed by the installation of climbing bolts to become an area of international importance for climbing (Howett K. pc). Stone quarries near the central lowland belt provide the main climbing opportunity for most of Scotland's climbers and reduce the pressure on natural mountain areas. Birnam Slate Quarry, Dunkeld, is described in the climbing guide by Cuthbertson (1983). The use of quarries for landfilling and leisure purposes is eroding the resource for climbers although in some instances co-operation has enabled climbing to be protected.

The British Sub-Aqua Federation regard the continued availability of Dorothea Quarry for winter training as a key priority since no other site offers such deep water in a sheltered accessible location in North Wales (Eccleston R. pc). The very deep water, reported to be about 150m in depth (Lovejoy and Partners 1988), has led to decompression accidents at Dorothea. Shallower quarries such as Fron Quarry, Nantlle are considered by some to be more suitable for diver training (Hughes T. pc).

4.5 Forestry and woodlands on slate workings

4.5.1 Unsuitable materials

The slate quarries of north west Wales, Cumbria and western Scotland produced waste which is very durable and has weathered little since tipping. The characteristics of

these wastes and the consequences for colonisation were described in sections 2.6 and 2.9 respectively. This waste material is unsuitable for large scale afforestation or woodland planting, although quarry sites have been partially planted when they lie within larger afforestation schemes. Natural colonisation has occurred on more sheltered slate waste tips, for example at Aberfoyle, Stirling and on tips where overburden was deposited with the waste, but to date this has not developed into mature woodland.

4.5.2 Commercial planting

The planting of forestry trees on slate waste for commercial purposes has been attempted on the softer Silurian slate wastes of Llangollen and Glyn Ceiriog, Clwyd. These wastes have decomposed on weathering to produce a relatively soft soil with a significant clay content, into which softwood species have been planted.

A survey of tree planting on various man-made sites was carried out by the Forestry Commission in 1974-5 (Broad 1979). This found that where slate had weathered sufficiently to provide a suitable planting medium, survival rates were usually high and crops were growing at a moderately fast rate. Table 4.2 shows the estimated yield classes of six species. Exposure and the degree of substrate weathering were regarded as important factors influencing the growth rates achieved.

Forestry Commission planting works at Beddgelert, the Lledr valley and Rhyd Ddu were reported by Crompton (1967). Survival and growth rates were very poor on the most inhospitable areas, but good where the slate had weathered. At Rhyd Ddu, planted in 1954, the Corsican Pine and Mountain Pine were found to be healthy but growing slowly after ten years. It was suggested that the use of soil beds and greater protection against livestock and rabbits were required for better results.

Commercial forestry is unlikely to play a significant part in further planting of slate waste tips because the establishment and growth rates would be uneconomic and the management and harvesting of timber on very steep land would be particularly difficult. Slate waste tips usually have side slopes of 38 - 42°, whereas the maximum slope for practical forestry is 27° (DOE 1989b) and lesser slopes would be selected where possible. Slate waste areas are expensive to fence effectively because of their size and shape, but the effective exclusion of livestock and rabbits is essential for the establishment of trees. The Forestry Commission has published a guide to the reclamation of disturbed land for forestry (Moffat and McNeill 1994).

4.5.3 Amenity woodlands

Revegetation trials and reclamation schemes have shown that woodlands can be established on slate waste if suitable planting techniques and site preparation are carried

Table 4.2 *Survey of forestry planting on slate waste*

Species	Estimated Yield Class slate waste	Typical Yield Class upland soils
Scots Pine	7 - 14	4 - 12
Corsican Pine	12[1]	6 - 14
Lodgepole Pine	6	4 - 10
Japanese Larch	8 - 12	4 - 14
Western Hemlock	12	12 - 24
Noble Fir	12 - 16	12 - 14

Notes:
[1] For Corsican Pine, Yield Class 12 equates to 11m tall at 25 years, 20m at 50 years.

SLATE WASTE DATA FROM BROAD (1979)

out. These treatments are described in section 5.2. The rates of growth produced by native pioneer species are insufficient to satisfy timber production objectives, but are sufficient for other objectives such as public amenity, recreation, wildlife conservation and landscape enhancement. The non-commercial species which can be established on slate wastes are generally better adapted to the conditions than commercial species, and are more suitable for these objectives. Once established, amenity woodlands require relatively little management provided that stockproof fencing is maintained, and can ultimately generate saleable products. Arfon Borough Council propose to establish a community woodland at the Allt Ddu reclamation scheme site, Llanberis (Hughes D. pc), and selective woodland establishment is likely to feature in sensitive, low-cost schemes for the rehabilitation of other slate workings.

4.6 Wildlife assessment and conservation in slate workings

4.6.1 Assessment of wildlife value

Site assessment is an essential first step in the evaluation and preparation of a management plan for a quarry regarded as valuable for nature conservation. It is also essential where a new quarry or an extension to workings is proposed, and where a new use is proposed for an abandoned site. Whether a site is abandoned or about to be developed, the principles of site assessment are the same. The Nature Conservancy Council published a method handbook for preliminary (Phase 1) field survey to identify and map habitats of wildlife importance (NCC 1990) (Box 4.1). Phase 1 surveys are the most appropriate for surveys of large areas of land, such as areas which may be subject to slate extraction proposals or to land surrounding existing sites, to place the wildlife resource of the site in context. A standard method for evaluating specific sites for their nature conservation and wildlife value was also developed by the Nature Conservancy Council (Ratcliffe 1977, NCC 1985), and has been further developed for the selection of sites of scientific interest (NCC 1989). The process of evaluating specific sites involves the systematic application of ten criteria (Box 4.2). In practice, site assessment will involve mapping different vegetation or habitat types and assigning values both to the different types of habitat and to the site as a whole. The National Vegetation Classification (NVC) has greatly aided the classification of vegetation and the determination of whether it is of local, regional or national interest.

Site assessments may show that only parts of a site are valuable for wildlife. The ranking of different areas on the basis of wildlife value will aid decisions about development, restoration or conservation. When making such decisions it should be noted that large sites are more valuable for wildlife than small ones, and that sites of value should be protected by buffer zones between them and areas of intensive land use.

4.6.2 Statutory protection

The Wildlife and Countryside Act 1981 is the primary legislation in Britain for the protection of wildlife. The main provisions of the Act are described in Department of the Environment Circular 32/81 (Welsh Office 50/81), SDD 3/82 and 31/86. The Act is generally administered by English Nature, Countryside Council for Wales and

Box 4.1 *Outline of method for Phase 1 survey*

Every parcel of land in the entire survey area is visited by a trained surveyor and the vegetation is mapped on to Ordnance Survey maps, usually at a scale of 1:10,000, in terms of some ninety specified habitat types, using standard colour codes. In practice much of the mapping can be carried out from public rights of way, using binoculars at relatively short ranges to identify the vegetation. Aerial photographs may also be useful, especially in urban and in upland areas, as an adjunct to ground survey.

The use of colour codes on the final habitat maps allows rapid visual assessment of the extent and distribution of different habitat types. Further information is gained from the use of dominant species codes within many habitat types, and from descriptive 'target notes' which give a brief account of particular areas of interest. The target notes are an essential part of Phase 1 survey and may provide the basis for selection of sites for Phase 2 survey and for decision-making in relation to conservation in the wider countryside.

Once mapped, the habitat areas are measured on the maps and statistics compiled on the extent and distribution of each habitat type. These statistics can then be held on computers.

The end products of a Phase 1 survey are:
(a)   habitat maps;
(b)   target notes;
(c)   statistics.

Ideally, the results should be supported by a descriptive and interpretative report. A descriptive summary for each Ordnance Survey map sheet has been found useful in some circumstances.

Scottish Natural Heritage (see Box 4.3).
Protection of species

Part I of the 1981 Act deals with the protection of individual plant and animal species. All birds, except for a few pest species, are protected from being killed, injured or taken from their nests, or having their eggs destroyed or their nests destroyed while in use. Certain rare species, listed in Schedule 1 of the Act, are given extra protection by special penalties. About 30 mammals, reptiles, amphibians and invertebrates, listed in Schedule 5, are given full protection under Section 9 of the Act. This protection includes prohibition of damage, destruction or obstruction of any structure or place used by the animal for shelter or protection, and disturbance of the animal while it is occupying this structure or place. Amongst the species protected are all species of bats, adders (*Vipera berus*), grass snakes (*Natrix*), and great crested newts (*Triturus cristatus*). Of these, bats are the most likely to be encountered at abandoned slate sites. If bats are encountered in a mine shaft, adit or old building and they have to be disturbed the government conservation agencies must be consulted. These agencies will offer advice on how best to proceed. Badgers (*Meles meles*) might also be encountered on old mine sites; they are fully protected in law by the Badgers Act 1990.

Ninety two species of vascular plants are included in Schedule 8 of the Act. It is an offence for any person intentionally to pick, uproot or destroy any of these wild plants. Protected species may be found at abandoned slate quarries or mines, or may form part of a habitat threatened by slate quarrying.

Protection of habitats

Part II of the Wildlife and Countryside Act 1981 deals with the conservation of habitats. A Site of Special Scientific Interest (SSSI) is defined by Section 18 of the Act as "any area of land which (in the opinion of the conservation agency) is of special interest by reason of any of its flora, fauna, or geological or physiographical features". Under the Town and Country Planning General Development Order 1988 (General Permitted Development (Scotland) Order 1992 in Scotland), Local Planning Authorities, or development control authorities in Scotland, are obliged to consult the organisations administering the Wildlife and Countryside Act 1981 when considering planning proposals that impinge on an SSSI. The owner of the land within an SSSI is obliged to consult the administering organisations before carrying out any changes in land use and management. For each SSSI a list of Potentially Damaging Operations (PDOs) is identified, eg the extraction of minerals, ploughing, planting, drainage, construction. If the owner or occupier of the land wishes to carry out any of these PDOs, four months' notice in writing must be given to the administering organisation. The organisation may consent to the operation or, if the operation is regarded as damaging, will approach the owner to negotiate a compromise,

Box 4.2 *Criteria for evaluating wildlife conservation value of sites*

Primary criteria

Size

In general, larger sites are more highly valued than smaller ones, all else being equal. Amongst aspects of size to be considered are the relative size of the site compared with sites of similar type, the extent of individual components of the site and whether the site is of sufficient size that small changes within will not lead to the loss of the site's interest.

Diversity

One of the most important site attributes is the number of communities and number of species. These are usually closely related, and in turn depend largely on a diversity of habitat. Diversity is sometimes related to habitat instability which will affect management prescriptions.

Naturalness

Ecosystems least modified by man tend to be rated more highly. However, the vast majority of sites of conservation interest have been influenced by man's activities to some extent. The degree and nature of this influence should be noted.

Rarity

Rarity is concerned with communities and habitats as well as individual species. The presence of one or more rare components on a site gives it higher value than another comparable site with no rarities.

Fragility

This reflects the degree of sensitivity of habitats, communities and species to environmental change. Fragile sites often represent ecosystems which are highly fragmented, dwindling or difficult to recreate.

Typicalness

Sites with examples of habitats which are characteristic of the ecosystem.

Secondary criteria

Recorded history, position in an ecological/geographical unit, potential value, intrinsic appeal.

---

which may involve compensation for any profit foregone by not carrying out the PDO. SSSI status does not guarantee protection; it ensures that conservation is considered during the planning process. Examples of SSSIs which will be destroyed or damaged by quarrying under old permissions were given by the RSNC Wildlife Partnership report 'Blasts from the Past' (1992). Kirkby slate quarry, Cumbria (Case Study 1) was one of these. Sites lying within National Nature Reserves, and many County Wildlife Trust Reserves, are better protected as they have management agreements in place with owners.

4.6.3   Protection in the planning system

Protection is also given to sites of wildlife value under struc-

ture and local plans. This protection will be strengthened under the provisions of the Planning and Compensation Act 1991 which allows for the inclusion of policies for protection for such sites in local authority development plans, and replaces a general "presumption in favour of development" by a presumption in favour of *development as defined by a local authority development plan.*

Guidance on nature conservation and planning is given in DOE Circular 106/77 (Welsh Office 150/77), DOE Circular 27/87 (Welsh Office 52/87) and DOE Circular 1/92 (Welsh Office 1/92) and in Scotland SOEnD Circular 13/91. Further planning policy guidance, which might supersede some or all of these circulars, was expected to be published in 1994.

Box 4.3 *Organisations responsible for nature conservation*

Statutory organisations

English Nature, Scottish Natural Heritage, Countryside Council for Wales. Formed from the Nature Conservancy Council, Countryside Commission (Scotland) and Countryside Commission (Wales), under the 1990 Environmental Protection Act.

These bodies have a statutory role to protect wildlife under the Wildlife and Countryside Act 1981, through the designation of National Nature Reserves and Sites of Special Scientific Interest, and through legislation protecting individual species. These organisations are statutory or discretionary consultees for many types of planning application.

Local government

Many councils have a conservation officer concerned with the implementation of policies for wildlife conservation.

Non-statutory organisations

RSNC, The Wildlife Trusts Partnership The UK's major voluntary organisation concerned with all aspects of wildlife protection. It consists of 47 Wildlife Trusts and 44 Urban Wildlife Groups which have local knowledge and manage some nature reserves on a voluntary basis.

Royal Society for the Protection of Birds, and many other specialist and local organisations covering badgers, bats, reptiles, woodlands and so on.

4.6.4 Consideration of reclamation proposals

The impact on wildlife of proposals for the reclamation of slate workings, for the establishment of new land uses or for the removal of slate waste as a secondary mineral is considered more rigorously in the 1990s than was the case in the 1970s and early 1980s. This is the result of growing official and public concern at the erosion of wildlife habitats, backed by the legislation described earlier in this section. A thorough evaluation of the wildlife value of any site where activity is proposed, and of the effects of that activity, should therefore be carried out as an integral part of the development of proposals. Schemes which minimise adverse impacts on wildlife, and incorporate measures to mitigate impacts and/or provide additional wildlife benefits, are more likely to receive public acceptance.

4.6.5 Access

The provision of access to areas of wildlife value should be considered on a site by site basis. Access by too many people, even if they are wildlife enthusiasts, to a site of value may damage it. At slate sites, trampling over scree may cause disturbance to species of value such as Choughs and so the level of access has to be carefully considered. In 1992, English Nature published its position statement on access and nature conservation which stated that "our objective is to maximise the benefits of nature conservation to people and in doing so, to ensure that the resource

itself is not degraded". Adoption of this objective on a site by site basis would be a valid approach, which should be integrated with considerations of the other resources eg geology or industrial archaeology within the site.

4.6.6 Examples of nature conservation

Few examples of specific nature conservation measures were found at slate workings although some operators stated that working procedures had been rescheduled to allow nesting birds to fledge, and that wildlife in non-operational areas of the quarry was not disturbed. Conservation is one objective of countryside management at the Tintagel cliff quarries SSSI and the Prince of Wales Quarry, north Cornwall and at Vivian Quarry, Llanberis, while at Aberfoyle Quarry, Stirling, the Scottish Wildlife Trust monitor a bat roost and the flora.

4.6.7 Conservation trails in slate workings

The potential for nature conservation trails varies from site to site. The older least disturbed sites offer the most wildlife interest, but in most quarries the nature conservation interest is probably secondary to other interests such as industrial archaeology or geology. Nevertheless, if trails through quarries are being constructed the opportunity should be taken to highlight those areas of wildlife interest or even to draw the walkers' attention to the difficulty plants have establishing on such waste. Hides and viewing platforms could also be constructed at those sites which are of ornithological interest. An integrated approach which encompasses wildlife has been taken at the Prince of Wales Quarry and engine house in Cornwall, where the mine trail and leaflet covers history, industrial archaeology and wildlife interests. At this site the basic requirements, listed in 'The Use of Land for Amenity Purposes' (LUC 1992), have been met. There is a car park, informal paths with disabled access where feasible, interpretive signs, safety fencing and an appropriate level of site management.

It is important that the provision of any facilities at quarry or countryside sites is carried out in a manner in keeping with the atmosphere of the place. The selection of alien materials, fussy details or an excessive scale of provision will destroy the sense of nature reclaiming the site from man.

Although slate quarries and workings do not have an intrinsically specialised flora or fauna, they may provide habitats not found on some surrounding land. Consequently such quarries add interest to the longer nature trails or country walks which are particularly popular in the Lake District and Snowdonia areas. Guide books and booklets which include descriptions of abandoned quarries, such as the booklet series 'Snowdonia and its Coast' (Johnson and Johnson (eds) 1992) help to widen public interest in the relationship between the countryside, its industries and nature.

4.6.8   The integration of land uses

The integration of interests can lead to conflicts which must be resolved by careful site planning. Uncontrolled public access to the Moel y Faen quarries, Horseshoe Pass, Clwyd has produced a conflict between those who explore the quarries and underground workings and the jackdaws, wheatears, choughs and bats (brown, long-eared, Daubenton's, Natterer's) which nest/hibernate and possibly breed there (Land Use Consultants, 1990). Elsewhere there may be conflict between the need to provide access to rock faces where geological formations are exposed, and the need to minimise disturbance of birds such as peregrines and barn owls which nest on inaccessible rock faces. Conflict may be reduced by confining users to separate areas, or by restricting access at nesting times. Legislation which protects certain species, notably bats, must be observed when any access or safety works are carried out. It is now standard practice when designing grilles to block adit or mine entrances, to use horizontal bars spaced to permit bat entry.

4.7   The conservation of geological features

4.7.1   Geological interest

Geological interest is maintained either by the conservation of existing exposures or, where there is a likelihood of new features of interest being exposed, by continued working in a sensitive manner. The rate of advance in most working quarries is such that new exposures are quickly disturbed, but the most valuable exposures are those in disused sites where safe public access can be gained. These exposures of minerals and formations provide material for teaching new geologists, and for research. Where the interest lies in mineral specimens or fossils, it has been argued that collection for teaching use elsewhere and the exploration for new specimens are important activities (Macdonald I. pc). Others have argued that specimen collecting is depleting a finite resource. Fossils within slate quarries are largely confined to the Devonian Caithness Flagstones, and so collection is not a major issue for geological conservation within most slate quarries.

Exposures created by active quarrying are rarely of a form which can be removed for safe preservation, and so only those which remain in the final landform are likely to be conserved. Where possible, conservation organisations will seek access for teaching groups or researchers at an early stage in the quarrying operation, and will encourage operators to provide a means of access to more remote exposures. Bodies such as English Nature will provide advice to operators to assist in the conservation of features (English Nature 1992).

4.7.2   Damaging activities

Activities which may disturb, damage or prevent access to geological features include afforestation, construction, insensitive reclamation, landfilling and neglect. Dense afforestation surrounding smaller quarries can physically prevent access, although public footpaths or quarry access routes can be used at many larger sites. The construction of forest roads can produce new shallow exposures of geological features. Landfilling, either with imported waste or slate waste, will obscure exposures unless the landform is designed to conserve features. This will inevitably reduce the volume available for landfilling and therefore affect the financial balance of the operation, but this may not be of consequence in non-commercial filling schemes such as land reclamation. Other uses of a site may be compatible with conservation if public access to exposures is retained. Slate waste tips can also be of value to geologists since they provide accessible specimens. The development of vegetation on quarry faces can slowly obscure exposures, but this vegetation can be managed where access is feasible. The management plan for a conservation site must therefore identify areas where geological exposures are to take precedence over natural recolonisation.

4.7.3   Statutory protection

The statutory protection of geological interest has much in common with the protection of wildlife interest. The conservation agencies English Nature, Countryside Council for Wales and Scottish Natural Heritage have an obligation to designate and protect Sites of Special Scientific Interest, under the Wildlife and Countryside Act 1981. This Act includes sites designated for their geological or geomorphological interest (Wright 1992). There are over 2,200 earth science SSSIs in Britain (English Nature, 1991), including a number of slate quarries, but as not all have been fully notified, the names are not reproduced here. The principal government guidance is still DOE Circular 27/87, although this may be superseded (4.6.3). Sections 31 and 32 (Development Planning and Development Control) of Circular 27/87 are of particular relevance. Section 31 stresses the need for local authorities to take the conservation of designated areas fully into account and include it in planning policies. Section 32 states "It is particularly important to take steps necessary to prevent valuable geological formations which are important for education and research from being obscured by dumping or tipping". The new PPG is likely to reiterate most of the points included in Circular 27/87 but has a new emphasis on 'sustainable development' and an enhanced role for local authority planners.

Planning and Compensation Act 1991

The Act has far reaching implications for all development, including waste disposal. The importance of conforming with development plans, including the Waste Local Plan, is stressed as a key determinant for approving planning applications. Geological heritage can therefore be given full consideration in these development plans. Local authority planners and industry should be made aware of the importance of protecting this heritage, where it is located and the options available to conserve and enhance sites.

4.7.4 Protection and planning

Most local authorities have policies protecting non-statutory sites of local or regional importance for wildlife conservation. Some authorities have policies protecting Regionally Important Geological/geomorphological Sites (RIGS). Most counties have RIGS groups, which aim to bring sites of local importance to the attention of local planners. These sites meet the needs of the local community for geological conservation and act as the backdrop to the SSSI network of sites of national importance. The non-statutory RIGS sites may achieve much the same protection in planning policies as local wildlife sites.

The existence of SSSI or RIGS status will alert those proposing new uses for slate working sites to the need to assess the value of that site and the likely impact of their proposals. Many sites have not yet been surveyed by RIGS groups and so the absence of such status should not be taken as confirming minimal value. The design of a scheme for reclamation, new use or mineral working should seek to conserve features of value where possible, for example by retaining rock or quarry faces in whole or part provided that the need for safety is satisfied. There is a need, therefore, for suitable geological expertise to be provided within the design team.

4.8 Industrial heritage conservation at slate workings

4.8.1 Objectives of conservation

The objectives of conserving the industrial heritage of slate working sites include:
- retaining sites and features for study in the future (preservation);
- presenting sites and features for current study and the interest of the public;
- conserving sites and features as part of the cultural landscape;
- providing a commercial tourist attraction.

Benefits of archaeological conservation

Archaeological features add to the character of an area, and are seen as a part of its cultural history. They enhance the education of children, residents and visitors, provide added attractions for tourists and encourage the conservation of valued landscapes.

4.8.2 Statutory protection

Statutory protection extends to Scheduled Ancient Monuments and to Listed Buildings. Examples of these are given in Table 4.3. The protection requires that the consent of the relevant statutory body must be obtained before any alteration or works may be carried out. The designation of sites and structures is reviewed periodically although the coverage of industrial sites and structures is regarded as inadequate (Grenter S. pc; Wakelin P. pc) and a programme of industrial site review is in progress. Further information is held by the national bodies and local authorities in the form of the National Monuments Register, and Sites and Monuments Registers. These list and describe a larger number of sites which are of interest, perhaps on a regional or local level if not nationally. The concept of 'national importance' is questionable in an industry which was necessarily localised. For example, is a typical Gwynedd slate mill of 'national' or just 'regional' importance? Are the typical small-scale features of the Lakeland slate areas such as the structures on Walna Scar of 'national' or just 'regional' importance? How are the cliff quarries of Cornwall to be classified? In Wales, the Gwynedd Archaeological Trust is carrying out a rapid survey to gather a basic database on the remains of the slate industry, so that site threats can be assessed more effectively.

Table 4.3 *Scheduled ancient monuments/listed buildings*

Site	National Grid Reference	Status
Pembrokeshire		
Porthgain harbour and brickworks (used waste from former slate quarry)	SM 814 324	SAM
Gwynedd		
Llanberis. Slate haulage table incline	SH 594 596	SAM
Dinorwig Quarry Barracks and 'A' incline	SH 586 603	SAM
Dinorwig Quarry workshop complex, waterwheel, Pelton wheel, table incline, water pipe and aqueduct	SH 585 602	SAM
Vivian Quarry incline, walls and associated structures	SH 586 606	SAM
Dorothea Quarry beam engine	SH 547 531	SAM
Dorothea Quarry pyramids, incline, mill and winding houses	SH 547 531	SAM
Gorseddau slate factory (Ynysypandy mill)	SH 550 433	SAM
Gelligrin slate quay (north bank)	SH 629 395	SAM
Pen yr Orsedd quarry Blondins and associated structures	SH 501 538	SAM
Coedyparc slate mill, Bethesda	SH 615 663	LB (II★)
Causeway at Dorothea Quarry, Nantlle	SH 495 538	LB ()
Cwm Machno incline and quarry housing	SH 750 471	LB ()
Highland		
Slate workers' boathouses, Ballachulish	NN 082 586	LB (B)

Key: SAM Scheduled Ancient Monument
        LB Listed Building (Grade I,II or II★ in England and Wales)
                                           (Grade A,B or C in Scotland)

Note:
There are no SAMs or Listed Buildings relating to the slate industry in England, and no SAMs relating to the industry in Scotland. Gelligrin slate quay is correctly named Tyddyn Isa quay (Williams M. pc)

SOURCES:

STRACHAN, 1991
K DAVIES, CCW, PC, 1992
P WAKELIN, CADW, PC, 1993
V COLLISON-OWEN, RCAHMS, PC, 1993.

4.8.3 Protection in the planning system

PPG 16: Archaeology and Planning (DOE 1990, WO 1991) sets out the means by which the planning system should recognise the importance of archaeology. The following extracts from Section A illustrate this.

"6. Archaeological remains should be seen as a finite, and non-renewable resource, in many cases highly fragile and vulnerable to damage and destruction. Appropriate management is therefore essential to ensure that they survive in good condition. In particular, care must be taken to ensure that archae-ological remains are not needlessly or thought-lessly destroyed. They can contain irreplaceable information about our past and the potential for an increase in future knowledge. They are part of our sense of national identity and are valuable both for their own sake and for their role in education, leisure and tourism."

"8. Where nationally important archaeological remains, whether scheduled or not, and their settings, are affected by proposed development there should be a presumption in favour of their physical preservation."

"18. The desirability of preserving an ancient monument and its setting is a material consideration in determining planning applications whether that monument is scheduled or not."

"14. But the key to the future of the great majority of archaeological sites and historic landscapes lies with local authorities, acting within the framework set by central government, in their various capacities as planning, education and recreational authorities, as well as with the owners of sites themselves. Appropriate planning policies in development plans and their implementation through development control will be especially important."

Unscheduled remains

It is significant that PPG16 paragraph 8 acknowledges that remains of national importance may not yet be scheduled (see 4.8.2). Industrial archaeologists still have a limited knowledge of the features on slate working sites, and believe that much more survey work is required before 'national importance' or 'regional importance' can properly be defined. The value of settings is also acknowledged, and this could be interpreted as applying to a whole site where working systems consisting of extraction, transport and processing are conserved.

A classic case is the complex in Cwm Ystradllyn, near Porthmadog, Gwynedd. The complex, which is based on the Gorseddau Quarry, has the classic lines of a Victorian slate quarry development, including a distinctive tramway system, a complete industrial village and an outstanding three-storeyed slate mill. The value of the whole complex is much greater than the sum of its parts.

The designation 'Area of Archaeological Importance' could be applied to such sites. Provision was made for designating AAIs in the 1979 Ancient Monuments and Archaeological Areas Act. Five areas have been so designated in England but there are none in Wales. However the legislation is not encouraging, there are no financial provisions made, and in Annexe 3 Paragraph 20 it is stated that no more AAIs should be designated until the success of PPG 16 has been monitored.

National Parks have now been given the responsibility for formulating Local Plans and, because their remit includes interpretation and recreation as well as conservation, it is recommended that they become consultees for industrial archaeology work within their boundaries (AIA 1991). This consultation could be carried out in partnership with the recognised authorities such as Cadw, English Heritage, RCHAM, the Archaeological Trusts and local authority Archaeological Officers.

4.8.4 The assessment of industrial archaeological value

PPG 16 states that advice on site assessment should be sought from the County Archaeological Officers in England and the Archaeological Trusts in Wales. These officers and organisations hold the Sites and Monuments Record (SMR) and have the expertise to make decisions of priority. Other relevant bodies are listed in Box 4.4. A good example of an evaluative approach is the Cornwall Archaeological Unit Report, 'Coastal Slate Quarries' (1990).

In Wales, detailed draft guidelines for assessment work on slate workings have been produced by the Welsh Industrial Archaeology Panel (1992). These guidelines are particularly relevant to proposals for large-scale disturbance such as land reclamation, slate waste removal or landfilling.

Underground remains

PPG 16 is concerned with the surface features of archaeology. In a mining and quarrying industry there are remains and evidence of techniques underground, with particular local variations which evolved primarily as a result of the differing dips in the slate strata. These features are not explicitly covered by PPG 16 and so planning authorities will wish to take further advice on underground remains. The umbrella organisation in this field is the National Association of Mining History Organisations (NAMHO).

4.8.5 Sites for future study

The preservation of intact sites for future study is not practised as a primary objective at any of the abandoned slate workings in Britain. Preservation would require considerable inputs to prevent revegetation, maintain structures and equipment and prevent deterioration in a manner which did not compromise the integrity of the remains.

4.8.6 Presentation of sites to the public

The prime function of a slate museum is to combine the preservation or conservation of sites and artifacts with their presentation to the public and to expert study. A museum should establish the highest standards of historical accuracy and material conservation, to complement the work of local authority archives and records offices in maintaining documents, for research and publication. This function is carried out at the Welsh Slate Museum, a part of the National Museum of Wales, but not at any other establishments relating to the slate industry in Britain. There is a prima facie case for a similar museum to represent the Cumbrian slate quarrying history, and for this function to be undertaken at suitable locations in Scotland and Cornwall. The Welsh Slate Museum forms part of Case Study 7.

Box 4.4 *Organisations responsible for industrial archaeology*

Statutory and advisory bodies

- English Heritage, Historic Scotland, Cadw: Welsh Historic Monuments
  Government bodies for the conservation of the built environment

- Royal Commissions of Ancient and Historic Monuments for England, Scotland and Wales
  Government bodies for the recording of archaeological and historic monuments or features

- Council for British Archaeology
  A representative body which provides advice to the government, and has regional advisory panels

- Scottish/Welsh Industrial Archaeology Panels
  Bodies of representatives from interested organisations that provide advice to the government and support research or survey work

- Local authority industrial archaeologists/archaeologists/records officers
  Usually based at County level. In England they hold the Sites and Monuments Records

- Archaeological Trusts
  Holders of the Sites and Monuments Record in Wales

- Countryside Commission, Scottish Heritage, Countryside Council for Wales, the National Park Authorities
  Have a role in the conservation and interpretation of cultural landscapes

Non-statutory bodies
- National Association of Mining History Organisations
- National Trust
- Welsh Mines Society
- Welsh Mines Preservation Trust
- The Lakeland Mines and Quarries Trust

4.8.7 Conservation within cultural landscapes

The conservation of sites within the cultural landscapes of National Parks or Heritage Coasts involves bodies such as the National Park Authorities, Heritage Coast department of the local authority, and landowners. Exceptionally a structure may be scheduled as an Ancient Monument and thus receive statutory protection, but this alone does not guarantee the conservation of that structure.

Each National Park that has sites of industrial archaeological interest has applied conservation management regimes. Predominantly this involves the consolidation of structures so that they can be 'left as they are' without deterioration. A successful operation is achieved when visitors are unaware that consolidation work has been undertaken on the structure. This work is usually followed by limited on-site interpretation which people can find for themselves.

Snowdonia National Park Authority have acquired a slate mill which, after consolidation work, is now a Guardianship Monument – Ynysypandy Slate Mill. Another site has been put forward for similar treatment – Hafodlas Quarry near Betws y Coed. Much research work has been carried out on a joint basis between the Snowdonia National Park Study Centre and Hull University, on the industrial archaeology of a number of sites, to assist with the presentation of these sites to the public. The Lake District and Pembrokeshire Coast National Parks have also produced some information on the industry in their respective areas.

Other bodies which carry out conservation work are local authorities and the National Trust. The partnership formed between the National Trust and the Cornwall Archaeological Unit of the County Planning Department to carry out a field survey on the cliff quarries is one example of good practice; another is the formation of the Easdale Folk Museum where the landowner and local inhabitants are creating a facility which has local support. In Gwynedd, the EC Leader Project is promoting 'cultural tourism' by producing information leaflets which include one entitled 'Industrial and Cultural Heritage'. This approach contributes towards a fuller understanding of the linkage between work, place and people.

4.8.8 Conservation and tourism

The commercial development of slate workings as tourist attractions ranges from those which present the history of slate working as the primary attraction, to those which make use of the setting for other unrelated activities. The conservation of industrial archaeology takes second place to the commercial objectives of the venture although the examples given in section 4.9 do retain a great deal of authenticity.

4.9 Tourism and educational uses of slate workings

4.9.1 Industrial tourism

Commercial tourism ventures which present industry, and particularly industrial history, to the paying public have generated increasing interest over the last decade, to the extent that the English Tourist Board designated 1993 as 'Year of Industrial Tourism'. The areas of slate working in Britain are all significantly dependent on tourism as a source of employment and income.

There are currently four slate-based commercial ventures in Wales, one in Cornwall and one on a much smaller scale on Easdale Island in Scotland. All four in Wales were established between 1972 and 1980 and so must be regarded as pioneers of industrial tourism. Two of these ventures are major employers and revenue earners in the town of Blaenau Ffestiniog where there are few other employment opportunities, and they introduce the industry and its history to a great number of visitors. Llechwedd Slate caverns attract approximately 250,000 visitors each year (Stevens 1987). Commercial activities may dilute or compromise historical accuracy; and some former quarry men are unhappy to see the places where colleagues or predecessors lost their lives being turned into a tourist 'playground'.

The Welsh and Cornish ventures all include an underground trip which, because it isolates the visitor from familiar reality, makes it possible to 'immerse' the visitor in the sights and sounds of mining or quarrying. This phenomenon, common to the constructed darkness of the Jorvik centre in York and many other sight and sound 'experiences', is also valuable for ventures which simply use the setting in which to stage themes not related to slate.

In Cumbria there is no such tourist venture connected with the slate workings although other industrial museums exist, such as the renovated bobbin mill at Newby Bridge. Kirkstone Quarries Limited do operate a slate and craft products showroom at Skelwith Bridge, but this is remote from the active quarry. Furness and Cartmel Tourism which promotes tourism in Southern Cumbria

believes that a visitor centre and tour/demonstration at the active Kirkby in Furness quarry would be successful, but this initiative has yet to be developed (Rogers pc). The Honister Pass quarries would be a spectacular location for a tourism development, but the visual and traffic problems of such a venture within the Lake District National Park would be extremely difficult to resolve.

4.9.2 Related attractions

A number of the narrow gauge steam railways of Wales owe their development to the slate industry, and maintain this link in their presentation to the public. The Ffestiniog Railway is perhaps the best known and most successful. Each year 400,000 tourists travel on the railway which once provided a direct link between the quarries of Blaenau Ffestiniog and the port of Porthmadog (Stevens 1987).

One other public attraction sited within a disused slate quarry is the National Centre for Alternative Technology at Llwyngwern quarry, near Machynlleth, Powys. Old quarry buildings have been renovated but the site's origins are clearly visible.

4.9.3 Education in slate workings

Education is one role of the Welsh Slate Museum, which welcomes school parties and similar formal groups as well as the general public. Many of the museum staff formerly worked in the industry and are able to demonstrate the skills of slate splitting, workshop skills and so on. Elsewhere, unmanned sites also have considerable education potential for industrial history, natural history and a variety of outdoor pursuits. Field work is a growing part of formal education and so the use of such sites is expected to grow. There is a need, therefore, to conserve sites in all slate-working areas to reduce the time and expense of travel and ensure that outdoor education is as widely available as possible.

4.10 Landscape assessment and conservation

4.10.1 The value of landscapes

Slate workings, and the cultural landscapes which surround many of them, form an important element in the landscape, contributing to or detracting from its value according to the viewer's perceptions. Where the contribution is positive in the view of a significant proportion of the population there is some justification for conserving that landscape and the slate working elements. Landscape conservation may be regarded as a positive use for such slate workings. The contribution of landscape value to the overall value of the National Parks, Areas of Outstanding Natural Beauty or Heritage Coasts in which slate workings are found was shown in section 2.12. Many features which make up the landscape have an intrinsic value, such as habitats for wildlife or items of archaeological interest. The value of the landscape as a whole is less easily defined but is a primary attraction within the areas mentioned. There is therefore a clear benefit in conserving the trees, woodlands and other vegetation, walls and hedges which make up the landscape. There is a benefit in retaining features which screen unsightly workings or soften the unnatural outline of tips in the manner described in section 2.12. Established vegetation, and particularly trees, can be effective in reducing the unnatural 'newness' of reclamation or redevelopment schemes, and can provide a buffer around slate waste extraction or similar processes.

4.10.2   The purpose of landscape assessment

Landscape assessments can be used to identify landscapes of particular value, for development control or protection purposes, and to identify the scope for enhancement as part of a positive programme of works. Landscape assessment methods are not yet commonly agreed, but the Countryside Commission analysed experience of landscape assessment methods and published guidance (Countryside Commission 1993), to facilitate 'consistency and comparison across the country'. This guidance was reviewed, and adapted to produce recommendations which are appropriate to the specific qualities and characteristics of slate working areas. The recommendations are presented in section 4.10.8 and Box 4.5.

4.10.3   Landscape assessment

In the past, there has been considerable confusion and debate over landscape assessment methods. During the 1970s, when the first round of county structure plans was in preparation, there was a strong emphasis on the use of supposedly objective, quantitative methods - usually aimed at identifying high quality landscapes for special protection. In reaction to these methods, which often failed to produce any real consensus as to landscape quality, more subjective, qualitative but systematic and structured methods were introduced in the 1980s. Again, landscape assessment was mainly used in the designation of special areas, such as Areas of Outstanding Natural Beauty.

Today, increasing attention is being given to sustainable development and to the conservation and enhancement of the countryside as a whole. There is renewed appreciation of the fact that the landscape is dynamic. Countryside agencies and local authorities are no longer acting in a purely defensive way but have begun to recognise the scope for creative action to guide and direct the evolution of the landscape. This can be achieved through planning and development control policies; through agriculture, forestry and land management practices; and through investment in new landscapes via schemes such as Community Forests and Countryside Stewardship.

Any such action should recognise and respect the distinctive 'sense of place' of different parts of our countryside.

Landscape assessment should help to identify cherished landscapes that are worth conserving, and at the same time highlight other areas in need of rehabilitation or improvement, so that resources may be targeted towards those areas where the greatest benefits can be achieved. At all stages, due account should be taken of public wishes and expectations. Landscape assessment is the essential starting point for creative and cost-effective landscape planning and management, and can also make an important contribution towards achieving sustainable development.

4.10.4   The meaning of landscape

'The term landscape refers primarily to the appearance of the land, including its shape, form and colours. It also reflects the way in which these various components combine to create specific patterns and pictures that are distinctive to particular localities. However, the landscape is not a purely visual phenomenon, because its character relies closely on its physiography and its history. Hence, in addition to the scenic or visual dimension of the landscape, there are a whole range of other dimensions, including geology, topography, soils, ecology, archaeology, landscape history, land use, architecture, and cultural associations. All of these factors have influenced the formation of the landscape, and continue to affect the way in which it is experienced and valued. Cherished landscapes can be said to have a natural beauty. This term embraces all the different dimensions of landscape listed above and also implies that the landscape is more than the sum of its parts.' (Countryside Commission 1993).

Sections 2.11-2.13 illustrate that a range of dimensions contribute to cherished slate-working landscapes, even though few observers would describe slate workings as having 'natural beauty'.

'Landscape is of fundamental importance in many ways. It is an essential part of our natural resource base. It contains valued evidence of earlier periods of human habitation and provides an environment for plant and animal communities. As human habitat it holds a special meaning for many people as the source of numerous experiences and memories.' (Countryside Commission 1993).

4.10.5   Basic principles of landscape assessment

'Landscape assessment is a general term for the process whereby a landscape is described, classified and evaluated. These three activities should be distinguished clearly from one another:

- **Landscape description** is the process of collecting and presenting information about the landscape in a systematic manner, and usually forms the initial basis for any landscape assessment;
- **Landscape classification** is a more analytical activity whereby the landscape is sorted into different types or units, each with a distinct, consistent and recognisable character;

- **Landscape evaluation** means attaching a value to a particular landscape, landscape type, or landscape feature, by reference to specified criteria. An evaluation generally should be based upon a prior classification' (Countryside Commission 1993). Table 4.4 presents a list of landscape evaluation criteria.

The distinction between landscape classification and landscape evaluation is particularly important. The former focuses on relatively objective recording and analysis of the intrinsic character of the landscape itself, but the latter includes a greater degree of subjective opinion and judgement as to the quality and value of the landscape. In practice though, all landscape assessments require a combination of objectivity and subjectivity. The key point is that the assessment process should be systematic and structured in order to gain acceptability.

4.10.6 Monetary evaluation of landscapes

There have been attempts to produce methods which derive a monetary value for 'resources' such as landscapes, with the purpose of drawing comparisons and allowing cost : benefit analyses between dissimilar factors. These techniques are not yet accepted or reliable since they depend on indirect assessment methods. The Department of Transport is supporting further research aimed primarily at the assessment of road schemes (DOT, 1992) but this is considered unlikely to become a widely used approach to landscape assessment outside road construction.

4.10.7 Slate working sites and their setting

In any area of slate workings, there are two elements of the landscape which need to be considered, namely the slate working sites, and the landscape setting of the workings. It is useful to look at these elements separately and then to weigh up their impact on each other. It is recommended that the landscape setting is examined first, so that the slate working sites can be seen in context.

4.10.8 A recommended approach to landscape assessment

As part of the Pilot Study conducted for this project, an approach to landscape assessment was derived from the various sources of guidance which are available. Following practical application and refinement, the approach presented in Figure 4.1 and Box 4.5 is now recommended for use in slate working areas:

- to assist in formulating a strategic approach to the area under study;

- to assist in defining sites which, on balance, make a positive contribution to the landscape and merit conservation;

Table 4.4 *Summary of criteria for landscape evaluation*

Landscape as a resource
- Rarity
- Representativeness/typicality

Scenic quality
- Combination of landscape elements
- Aesthetic quality
- Intangible qualities   - sense of place
                          - 'habitat' theory

Preference
- Public preference
- Informed consensus

Special values
- Wild land qualities
- Cultural associations
- Special heritage interests   - wildlife
                                - archaeology/history
                                - geology/geomorphology

THIS SUMMARY WAS PREPARED FOR THE FORMER COUNTRYSIDE COMMISSION FOR SCOTLAND (LAND USE CONSULTANTS 1991).

Figure 4.1   Recommended landscape assessment process

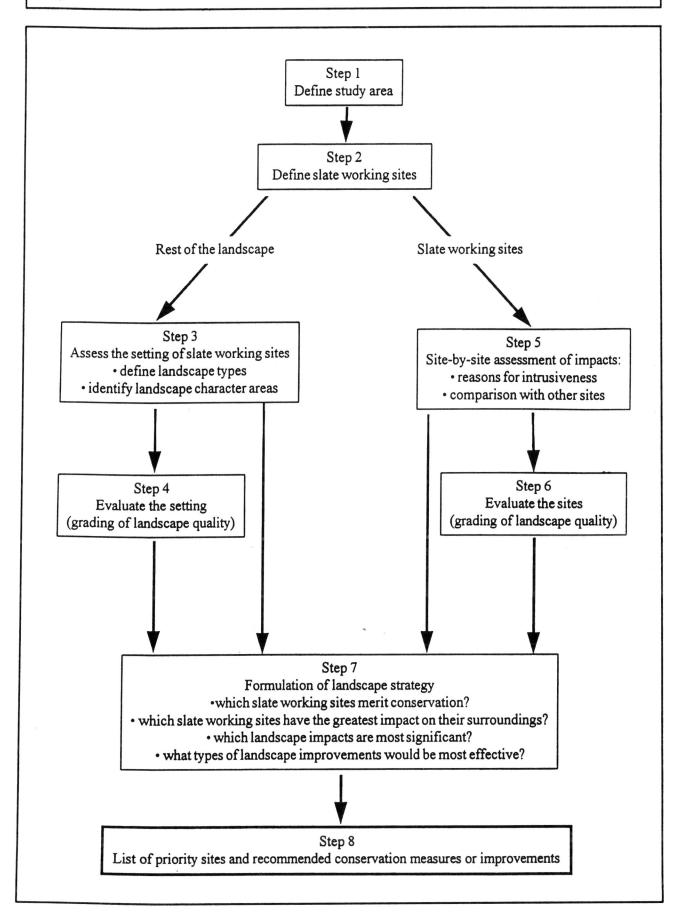

Box 4.5 *Recommended approach to the assessment of slate-working landscapes*

Step 1   Define the study area

Define the area to be assessed by identifying the visual catchment of all the slate-working sites in the area being considered. The visual catchment is that area which can be seen from, or provides a view of, any part of a site. In hilly or mountainous terrain long-distance views are frequently possible and discretion should be used to define a limit to the area to be assessed.

Step 2   Define the slate-working sites

Identify the boundaries of the areas disturbed by slate working, using Ordnance Survey maps and aerial photographs. These areas will be treated separately from the rest of the landscape ie the setting, in the remainder of the landscape assessment. The sites will form one landscape type ie areas of the landscape which exhibit consistent broad characteristics. Visually distinct types of workings could form separate landscape types.

Step 3   Assess the setting

The setting is classified by dividing it into landscape types, and then sub-dividing it into landscape character areas ie unique areas of definable, distinct character. The primary factors examined in the assessment of landscape types and character areas are landform and land use, but ecological, historical and cultural associations may also be important. Landscape types and character areas are defined by field work and desk study.

Step 4   Evaluate the setting

Once areas of distinct character have been defined, the landscape quality of the setting of each slate-working site may be considered. A simple, subjective quality grading such as very high, high, medium, low and very low is recommended. This grading indicates the significance of the impact of each site on its surroundings.

Step 5   Site-by-site assessment of impacts

Examine each site in a systematic and structured manner, in order to rationalise the reasons for the intrusiveness of the site in the landscape. Compare each site with the others to assess their relative intrusiveness. The intrusiveness of a slate-working site is a subjective and sometimes contentious matter, linked to public awareness of industrial history. A check-list has proved to be a useful way of recording site observations. An example is reproduced in Figure 4.2.

Step 6   Evaluate the sites

A simple quality grading of the sites, indicating their relative impact on the landscape, may be derived from steps 4 and 5. The quality grading is one component of step 7.

Step 7   Formulation of a landscape strategy

Having defined the landscape character of the surroundings and commented on their intrinsic quality, and having studied the slate-working sites, it is possible to decide:
- which sites make a positive contribution and justify conservation?
- which sites should be the priority for treatment? ie which sites have the greatest negative impact for the greatest number of people?
- what characteristics of these sites have the greatest visual impact and what could be done to amend these characteristics?

Step 8   List the priority sites

From these decisions a strategy, consisting of a list of priority sites and recommended conservation measures or improvements for those sites, can be drawn up.

- to assist in defining sites which, on balance, have a negative effect on the landscape and in defining what type of landscape improvements would be most effective, if any;
- to assist in defining priority sites, or parts of sites, where such improvements would be most effective.

The published guidance (Countryside Commission 1993) gives detailed advice on the methods and techniques involved.

4.10.9 Landscape conservation in the planning system

Any proposed development which requires planning permission is dealt with as set out in section 2.4. Local Planning Authorities may designate Areas of Special Landscape Value, or similar designations, within which they apply particular landscape scrutiny to proposals. In nationally designated Areas of Outstanding Natural Beauty, or Heritage Coasts, or National Parks, the planning authority will apply even greater scrutiny to proposals in order to conserve the landscape, and will consult the Countryside Commission or Countryside Council for Wales. These designations do not at present apply within Scotland, although Scottish Natural Heritage would be consulted on sensitive landscape issues. The designation of Tree Preservation Orders is sometimes used as a tool for landscape conservation, and would be a material factor in the scrutiny of proposals. Designations made primarily for the conservation of wildlife or geological features, eg SSSI, may often have a landscape conservation benefit.

4.10.10 Landscape management and conservation schemes

Various schemes exist to encourage landowners to maintain and conserve the landscape and its features, and to allow access. Many involve some grant payment towards conservation works. Organisations such as those listed in Box 4.3, together with local authorities acting as agents, administer and promote these schemes. These organisations also carry out conservation works directly on land they own, or on private land by agreement. The main slate working areas of Wales are, however, excluded from the Snowdonia National Park and to date very little such work has been carried out on abandoned slate workings (Rendell T. pc).

4.10.11 The effects of land reclamation

Large scale land reclamation, removal of slate waste or other change in the landscape has the potential to remove much of the landscape character of a site and its surroundings. There is a need to carry out a landscape assessment for the purpose of description and analysis (see 4.9.5) to identify which features and patterns create the landscape character, so that appropriate design decisions are made and the result is not alien to the surrounding landscape. This may involve the scale and pattern of enclosures, the scale and shapes of woodlands, the slopes and changes of slope, the types of vegetation and so on. The materials used to construct features such as walls, stiles and fences should respect those in the vicinity. In many areas slate was, at one time, used as a walling stone and building material, and a number of particular local styles may be seen in slate-working areas. Traditional walling often accompanies road widening schemes, and slate waste is supplied for this purpose in Cornwall (Hamilton G. pc). The Welsh Office encourage the use of sympathetic design detailing in their guide 'Roads in Upland Areas' (Welsh Office 1990). This stresses the need to recognise the subtle local variations in materials and methods if matching detailing is intended. Such attention to detail may well add slightly to the cost of a finished scheme but makes a great difference to the appearance.

Figure 4.2 Checklist for use in site-by-site assessment

Site name ...Rhosydd Quarry (NB upper area)... Grid ref ...SH 66 6 4 5 5...

Date of observation ...16-11-9?... Observer ...AJ...

GENERAL VISUAL CHARACTERISTICS

The Site A patchy area of small/medium tips & quarry holes (one very large) - unusual in this area. Tips are low and old, partly vegetated (grass, moss, lichen). Rock is more shaley than in lower area & different from lower area & same site.

The setting Landscape type (High elevation (c500m) moorland, (Molinia & sphagnum) - nearly level area on N. slope of Moelwyn mountains)

Character areas (not detailed in this study)

Context of site The end of a sequence of sites, starting at Tan-y-grisiau, rising out of the Cwm. Unlike all the others, this area is unrelated to the Cwm.

VISIBILITY

Seen from where/by whom? Seen only by mountain walkers on hills around & walking through site. Three public footpaths traverse the site. Well used paths - but not a Snowdonia 'honey-pot'.

Distance of views — mixed : long views (500-1000m) from hill tops, down to paths through the site.

VISUAL PROMINENCE

Size moderate/small | **Scale, in context** small

Mellowing Yes - v.much so - grass, moss, lichen. Plus different rock | **Climate/aspect** High elevation - often in cloud

Colour, texture, landform, in context Tips are low & flat in a level area. Holes not prominent because of nearly level topography. Vegetation & rock colour reduce prominence

Eyecatching features/skylines Not strong, cf. other sites.

Other factors of note Ruined buildings now low-level. Quarry holes striking but not v. visible. General variability of site reduces impact.

POTENTIAL FOR LANDSCAPE IMPROVEMENT

Visibility or prominence factors which could be treated Holes could, in theory, be filled with nearby waste - but site is of low prominence compared to most. Practical problems of any treatment would be severe in this location.

Priority for treatment - compared to other sites Low priority - Recommend no action.

QUALITATIVE ASSESSMENT

(tick one)

High quality .... Poor/damaged .... ✓

Average .... Severely damaged ....

Comments A modest-sized site, not very prominent in the landscape; interesting to walk through "unspoiled" by recent working.

OTHER COMMENTS OR NOTES

It would appear that underground access occurs here via quarry hole.

Public safety (unfenced holes, sheer sides) is an issue - especially in low visibility conditions. But fencing would be obtrusive

5 Reclamation and Rehabilitation

5.1 A review of reclamation schemes

5.1.1 Land reclamation programmes

Very few slate working sites have been restored or reclaimed by the industry as a requirement of planning permission conditions, but in future the industry is likely to have to take increasing responsibility for the treatment of its current sites. The Department of the Environment's Derelict Land Grant Advice Note 1 of May 1991 states that it attaches high priority to the reclamation of derelict land and that current policies are designed to encourage the return of derelict land to beneficial use as soon as possible. Land reclamation programmes are long established in England, Wales and Scotland under the sponsorship of the Department of the Environment, the Welsh Development Agency and the Scottish Development Agency, subsequently Scottish Enterprise, respectively. Grant-aided land reclamation schemes for disused slate workings have been carried out at one site in Scotland, fourteen sites in Wales, and one site in England, in the period 1972-1992. English Partnerships has now been established to operate a unified grant regime in England, in place of the regime operated by the DOE. The grant-aid schemes operated by each of these bodies are outlined in section 5.5.

Appendix 3 presents a summary of some examples which illustrate the range of works carried out in these land reclamation schemes. The case studies presented in Annexe 1 give more detail of selected schemes.

5.1.2 Scheme objectives

The objectives of the reclamation schemes considered during this research have been set out on a site by site basis, although each scheme fitted into the overall priorities of the respective grant-aid programmes. The scope and content of each scheme was determined by the local authority body leading the scheme but the funding guidelines and technical requirements of the grant-aiding body also influenced the schemes described in Appendix 3. All the schemes carried out to date have largely or completely met the objectives that were set for them. In the light of current attitudes, approaches and techniques it is possible to be critical of some design details, of the lack of concern generally for the conservation of features which today would be regarded as valuable, and of the selection of objectives for the schemes. Not withstanding this, the schemes tackled major problems and much has been learned from the experience.

5.1.3 Scheme funding

Many of the schemes described in Appendix 3 and the Case Studies were constrained by the availability of funds. The grant-aid covered essential engineering and revegetation works but was not extended to the provision of some details, a point which might be criticised today. For example, the larger schemes such as Talysarn and Allt Ddu created large open grassed spaces which are now considered to be out of keeping with their surroundings. At Talysarn it was proposed at the outset to construct drystone walls of slate to match those nearby, but consideration of cost meant that the cheaper option of post and wire fencing was adopted (Richards I. pc). Particular sensitivity is required in the design of large scale reshaping works, which cannot easily be altered after completion, and in the attention to detail at the edges of a scheme where the works adjoin the surrounding public or private properties. Details such as entrances, gates and stiles, walls and fences are those noticed at close range by neighbours and visitors who will appreciate their quality, or lack of it. These details generally form a relatively small part of the overall scheme cost, and yet form a major part of what most people see of the finished product. Recent and current reclamation projects elsewhere in Wales have paid much greater attention to these details as part of a new emphasis on improving the landscape, under the WDA's "Landscape Wales" reclamation efforts. Whilst the landforms created by the early reclamation schemes can only be changed at great expense the opportunity remains for the addition of further landscape details to these sites when the resources are available.

5.1.4 Cost of works

Where scheme cost information has been available it has been recorded in Appendix 3, together with a description of the works involved. The scheme costs ranged from £5,000/ha to £125,000/ha (Table 5.1). It is not considered helpful to draw these together to estimate an average cost per hectare or cost per cubic metre of slate waste, since every scheme has had its own objectives and constraints, initial problems and design solutions. The cost of a scheme reflects not only the quantities of earthmoving and construction works, but also the characteristics of the site. Schemes carried out in close proximity to the public were typically more expensive since greater attention was given to minimising nuisance and to providing boundary treatments which benefited all parties.

The principal elements contributing to the cost of the reclamation schemes were:
- land purchase;
- scheme design and administration;
- clearance of structures and debris;
- pumping of flooded quarries;
- diversion of surface waters;
- excavation, transport and deposition of slate waste;
- compaction of fill;
- trimming of regraded surfaces;
- surface crushing, preparation and seeding;
- provision of surface water drainage;
- provision of infrastructure for new uses;
- safety barriers, fences and walls;
- planting and stockproof fencing;
- maintenance of vegetation.

5.2 Reclamation techniques

5.2.1 Introduction

The techniques used in these reclamation schemes are described in this section, and the lessons which were learned are brought out. This section also describes newer techniques for revegetation which were not available or wholly appropriate for many of the schemes carried out. Many of these techniques are also applicable to restoration work carried out by a mineral operator. Section 5.3 discusses the approach to site assessment, scheme planning and design which would be applied to sites today, showing some contrasts and some similarities with the approach to earlier schemes.

5.2.2 Excavation and movement of slate waste

Most of the schemes described relied on the large scale movement of slate waste. Conventional civil engineering techniques were blended with quarrying methods to deal with the extremely coarse material encountered.

Slate waste is generally too coarse to be excavated by box-scraper and so excavator and dumptruck methods were used in most cases. At Glan y Don however, large quantities of mill fines were uncovered, which were suitable for movement by box scraper (Richards I. pc). Movement over short distances and minor regrading was feasible using bulldozers which also had the effect of breaking down the slate into small pieces. It was found that schemes requiring slate waste to be moved downhill were significantly easier than those involving uphill haulage. For example, a significant cost saving was achieved in the construction of the Dinorwig Power Station by moving slate waste downhill to create a working platform in the lake bottom rather than moving it uphill to other potential disposal areas. Vehicle tyres were rapidly worn or punctured by the sharp-edged slate fragments. Contractors overcame this by adopting chain protection for loading shovels and by selecting fine slate to surface the haul roads. The size of plant which could be used was often restricted by very narrow or winding roads leading to sites or connecting excavation and fill areas. In some instances equipment was dismantled, transported in pieces and reassembled on site. The process was reversed on completion of the schemes.

Due to the hardness of the slate and its free-draining nature, the movement and regrading of slate waste was able to continue throughout the wettest winter weather. This factor contributed significantly to the low contract rates since contractors were able to avoid unproductive standing time. As earthmoving typically accounted for 75% of the work in the larger reclamation schemes, the savings were substantial.

5.2.3 Slate waste as a fill material

Slate waste is inherently suitable as backfill for quarry holes and as fill to raise low lying ground, since it is inert and resistant to degradation. Settlement has not been observed on sites where land has been filled for a new purpose even where fill depths of 12-15m were required, as for Corris Craft Centre, though at Corris building did not take place for some years after the fill was placed. At Blaenau Ffestiniog, houses were built on fill within 3 years without any problems due to settlement, the fill being 3 to 4 metres in depth. Some random cracking occurred in the carriageway of the A487(T) in the 12 months following completion. This cracking was attributed to settlement in uncompacted old slate waste lying beneath the engineered fill, caused by the superimposed load of the new embankment. The cracks were sealed and no further problems have been reported.

The main shortcoming with slate as a fill material is its platiness which makes it difficult to place in layers. Care was taken in the schemes described to ensure that the earthmoving machinery placing the material in a fill area was of an adequate size and capacity to deal with the largest pieces of slate being delivered. In development areas fill was placed in 600mm thick layers in order to maximise the in-situ crushing effected by the earthmoving machinery. Care was also taken in these schemes to ensure that:
- scrap cars and the like were removed before filling;
- standing water was pumped out before filling;
- slate fill was tipped, spread and compacted in layers rather than loose tipped;
- surface layers were compacted;
- areas to be filled were dug out where necessary to give a minimum of 1.5-2m of compacted slate fill.

5.2.4 Specification for earthworks

Civil engineering contracts for all the schemes considered were based on the industry standard 'Specification for Road and Bridge Works', amended slightly for surface finishing treatment. Large pieces of slate were excluded from the general fill, and compaction of the surface layers was achieved by excluding material greater than 0.02m³ from within 0.6m of the surface, and applying ten or more passes of a grid roller. The specification for Ballachulish required fill to be placed in layers not exceeding 225mm and compacted with ten or more passes of a vibrating roller of 1800-2300 kg mass per metre width (Sparkes pc).

Unlike finer waste materials such as colliery shale, slate waste does not bulk up when excavated and replaced. In practice, with typical compactive effort applied to the slate waste during filling, a ratio of 4 volumes excavated to 3 volumes tipped and compacted was commonly achieved. This reduction factor was taken into account in scheme designs (Richards I. pc) and additional provision was made for variation in the ratio actually achieved, in the form of balancing areas for which the final contours could be adjusted as work progressed.

5.2.5   Design of drainage systems

Although slate waste is very free draining, scheme designs had to provide for the drainage of water issuing from the base or lower regions of tips or filled areas, and the surface run-off which occurs in periods of heavy rainfall. The compacted slate waste surface of a regraded or filled area consists of many overlapping platy fragments. As a result the drainage path for surface water is a series of more or less horizontal laminar channels and vertical drainage can be slow. The reclamation sites are located in the wet western uplands of Britain where the heaviest rain often occurs after prolonged rainfall, by which time the catchment is saturated. The typical drainage design parameter used is that of a 100 year return storm falling on a saturated or frozen, snow-laden catchment producing a run-off factor of 1. The drainage channels must have the capacity to cope with large volumes and high flow rates, even though they carry only a trickle for much of the year.

5.2.6   Establishment of grass

The majority of reclamation schemes have not used topsoil or subsoil as a growing medium for grass since large quantities of soil are not locally available in the upland slate areas. Surface preparation techniques have been developed, in which slate waste is crushed by repeated passes of grid rollers and sheep's foot rollers. This produces a thin layer of fragments, fine enough to allow the establishment of grass. An alternative method, used at Braichgoch, Corris, was to crush slate waste in mobile quarry crushing and screening plant to produce road subbase, drainage backfill and fines of 1½" (40mm) down. The fines were used as a surface dressing on the new slopes, which were as steep as 1 in 2 in places, in preparation for seeding. These slopes were too steep to be rolled (Richards I. pc).

The principal method of grass seeding used was initially developed in colliery spoil reclamation. The seed mixture was broadcast over the crushed slate surface, and top dressed with broiler house litter at 10t/ha. The litter acted as an organic mulch which was resistant to high rainfall, and provided nitrogen and phosphate. The nitrogen was available to the grass over a short period which promoted very rapid growth over one season. Where a productive grazing sward was the objective this technique was appropriate, and if the required maintenance fertilisers had been applied it would have been successful. However, the rapid grass growth tended to exclude the clover from the sward. Where clover did establish, as at Talysarn, the lack of further phosphate applications coupled with excessive grazing led to the death of the clover. As a result, there was very little long term nitrogen supply, the initial flush of grass could not be sustained and the productive species tended to become moribund and to die out. Since maintenance has generally been lacking on these reclamation schemes it has been necessary to consider other objectives for revegetation such as low-maintenance, low-productivity swards for recreational use. These objectives generally lead to a sward which is slower to establish and much less vigorous in its early appearance, which can be in conflict with the expectations of a rapid 'greening' and instant landscape sometimes held by the public.

Hydraulic seeding has been used where slopes were not accessible to conventional plant, but this technique is relatively expensive and so has not been widely used. At Braichgoch the hillside slopes, which ranged from 30-40°, were dressed and seeded by hand after the removal of the slate waste. The application of maintenance fertilisers to such steep slopes would be difficult and has not been carried out.

A research project into low cost techniques for the reclamation of derelict land was commissioned by the Welsh Development Agency in 1982. Trials of methods to establish grass without regrading slate waste tips were conducted (Robinson Jones Partnership 1987). It was concluded that the surfaces of most slate waste tips do not contain sufficient fine-grained material to allow grass to establish. Only where conditions were favourable, ie where shade from the sun reduced the intensity of desiccation or where pockets of organic material had accumulated, would grass establish from seed. On the whole, it was considered that reliable grass establishment required substantial modification of the waste tip surface.

5.2.7   Selection of grass species

The seed mixtures used for early schemes were based on perennial ryegrass, red fescue and clover. Rapid establishment was a priority and, where grazing was the intended land use, species giving palatable and productive growth were preferred. These species are only vigorous and productive in conditions of adequate fertility and freedom from drought. Such conditions are very difficult to create and sustain on a coarse, free-draining substrate consisting of slate fragments. Consequently, the growth of the swards declined rapidly and the grass failed to persist. Studies of the natural colonisation of slate waste tips and similar nutrient-poor, drought-susceptible substrates showed that slow-growing native species of grass were able to withstand the stressed conditions and persist, albeit with a low productivity, without significant management inputs provided that grazing was restricted (Robinson Jones Partnership 1987). Later schemes introduced a wider range of mixtures, incorporating these native species and varieties, designed to tolerate minimal maintenance and very low nutrient status in the substrate.

5.2.8   Experience of tree planting

Tree and shrub planting has formed a part of most reclamation schemes. The techniques developed as experience was gained. In an early scheme, Coed Madog at Talysarn, trees were planted in pits of imported topsoil, the pits being large enough to accommodate the tree roots. This method produced limited success since the tree roots tended not to extend into the surrounding regraded slate waste and the topsoil was prone to drying out.

At Glan y Don, Blaenau Ffestiniog and at Braichgoch, Corris, imported subsoil was placed in beds 100-150mm deep. Trees ranging from transplants (300mm tall) to light standards (1.8-2.1 m tall) were planted in these beds. The soil and underlying slate were well mixed in the root zone. Although some species, notably ash and oak at Braichgoch, have grown very slowly, in general the trees did establish and have produced dense blocks over 10m tall. The Ballachulish scheme in western Scotland included the planting of 85,000 transplants (450-600mm tall) and feathered trees (1.8-2.1m tall) into beds of imported topsoil, 100mm deep. Very high establishment rates were followed by rapid growth which has been maintained.

5.2.9   Planting without soil

More recent schemes have included tree planting without topsoil or subsoil. At Allt Ddu, Llanberis, selected slate waste from the tips was crushed and screened on site to produce fines in the range 25mm to dust with a maximum of 5% exceeding 10mm. A 250mm layer was placed over 4 ha as a soil substitute, and trees were planted in pits backfilled with 1 part planting compost to 2 parts slate fines to which fertiliser was added. Polythene sheet mulch was applied to reduce moisture loss. Despite losses due to feral goats and extreme wind rocking, the results were generally good. A similar planting method was used at Abercwmeiddau, Corris, where sufficient fine slate was obtained by selection during the excavation of the tip without the need for screening.

One other planting technique which has been developed for tree planting on undisturbed and coarse slate wastes, is 'pocket planting', shown in Figure 5.1. The method was first reported following experimental work by Sheldon (1975), and was refined in further trials carried out as one element of research for the Welsh Development Agency (Robinson Jones Partnership 1987).

The method was applied to small planting areas at the Fotty Tip scheme, Blaenau Ffestiniog with moderate success.

5.2.10   Selection of tree species

The WDA research demonstrated that species which naturally colonise slate waste tips, such as birch, rowan and goat willow, could be established in undisturbed slate waste. These species are able to persist despite the stresses of drought, lack of nutrients, and exposure. Alder, which is uncommon in the vicinity of most slate waste tips, was also shown to grow rapidly due to its nitrogen-fixing ability (Blunt 1991). A maintenance fertiliser application promoted growth rates 15% greater than those achieved by unfertilised trees, but even without such fertiliser, tree growth continued. An inspection early in 1993 showed that the trial plots were continuing to develop into woodland eight years after the last fertiliser application, and that accumulating leaf litter was beginning to form an organic layer. This layer was being colonised by grass and herbaceous species.

5.2.11   Tree establishment from seed

A number of researchers (Blunt 1991; Al-Gosaibi 1985; Luke and MacPherson 1983) have experimented with methods to establish trees and shrubs by sowing treated seed directly onto slate waste and mineral shale surfaces, and applying organic mulches to protect the seed from desiccation. The success achieved on finer materials such as colliery spoil was not matched in trials on slate waste. It was concluded (Blunt 1991) that unless the slate substrate was sufficiently fine textured to retain some water yet sufficiently open and fissured to allow some natural burial of the seed, then economic rates of mulch were insufficient to protect germinating seed from desiccation. Examples of extensive natural colonisation of crushed slate surfaces were noted during the fieldwork for this review at Allt Ddu, Llanberis; Y Glyn, Llanberis; Braichgoch, Corris; Hodge Close, Tilberthwaite and the Centre for Alternative Technology, near Machynlleth.

A dense grass sward or moss surface is known to inhibit or prevent the successful establishment of incoming tree seeds (Finegan et al, 1983). In the examples noted in the previous paragraph, colonisation was in areas of thin, sparse grass or areas without grass. This suggests that where the potential for desirable natural colonisation exists, initial grass treatment should be directed towards an open sward of low vigour, although this may not satisfy some expectations of a rapid initial green cover.

5.2.12   The choice of vegetation types

The choice of vegetation types for use in reclamation

schemes is influenced by the intended use of the site, and by site factors including:

- compatibility with surrounding vegetation and the landscape;
- tolerance of climatic and substrate conditions;
- performance under conditions experienced after reclamation;
- effects on the stability of slopes and structures.

The extreme exposure of many upland slate working sites is a severe constraint on the establishment and growth of trees and shrubs, and in areas where sheep graze freely it is unlikely that trees will survive. The choice of vegetation types should be compatible with the land management regime operating over the surrounding area and the site. Once the vegetation types have been chosen, the appropriate species can be selected taking account of the factors and experience described in the preceding sections.

5.2.13   The need for continued management

In the majority of the schemes described in Appendix 3 and the Case Studies, vegetation management ceased at the end of the contract maintenance period lasting for 1, 2 or latterly, 3 years. A great contrast may be seen between unmanaged sites such as Glan y Don, Talysarn and Allt Ddu where uncontrolled, unauthorised grazing and a lack of any maintenance fertiliser have combined to reduce the vegetation cover to a sparse grass sward dom-

inated by moss; and managed sites such as Ballachulish and Cwm Penmachno where the intensity of grazing has been balanced by initial and sustained fertiliser or manure applications to maintain a productive grass-dominated sward. This is one of the principal conclusions to be drawn from this review of reclamation schemes.

Regraded and revegetated slate waste, in common with other derelict land substrates derived from inert mineral wastes, cannot be regarded as 'reclaimed' on completion of the engineering works. This has been recognised in modern mineral extraction, for which planning conditions can require an aftercare programme of up to 5 years (MPG7) after soils have been replaced. Inert mineral wastes require regular and continuing care to ensure that a soil develops. Only then can the site be treated in the same manner as the surrounding grazing land. If the intensity of grazing exceeds the intensity of management inputs, principally fertiliser, then the sward will regress and die out just as it will in grazing land on natural soils. Derelict Land Grant Advice Note 1 emphasises the need to guarantee the maintenance of land after reclamation. Current Welsh Development Agency requirements for new schemes call for a management strategy and proposals as part of the scheme design (Griffiths DG. pc).

A reclamation project can only be fully successful if the vegetation which is established is sustainable under the conditions of use and maintenance which follow the ini-

Table 5.1 *Costs of land reclamation schemes described in Case Studies and Appendix 3*

Scheme	Case Study/ Appendix 3 No	Derelict area [1] ha	Scheme cost [2] £	Cost/ha £
Allt Ddu	CS4	24	1,600,000	66,666
Ballachulish	CS5	25	2,582,300	103,292
Coed Madog	3	8.7	162,200	18,646
Braichgoch	5	15.3	1,596,875	104,371
Glan y Don	6	11.9	1,497,436	125,835
Talysarn	7	17	1,291,156	75,950
Y Glyn	8	2.5	62,275	24,910
Cwm Penmachno	14	4.9	26,071	5,321
Abercwmeiddaw	15	8	550,000	68,750

Details of the schemes are given in Case Studies 4 and 5, and in Appendix 3
1. The derelict area is often less than the completed scheme area.
2. Scheme costs have been indexed to a base year, 1990.

Figure 5.1 'Pocket' planting method

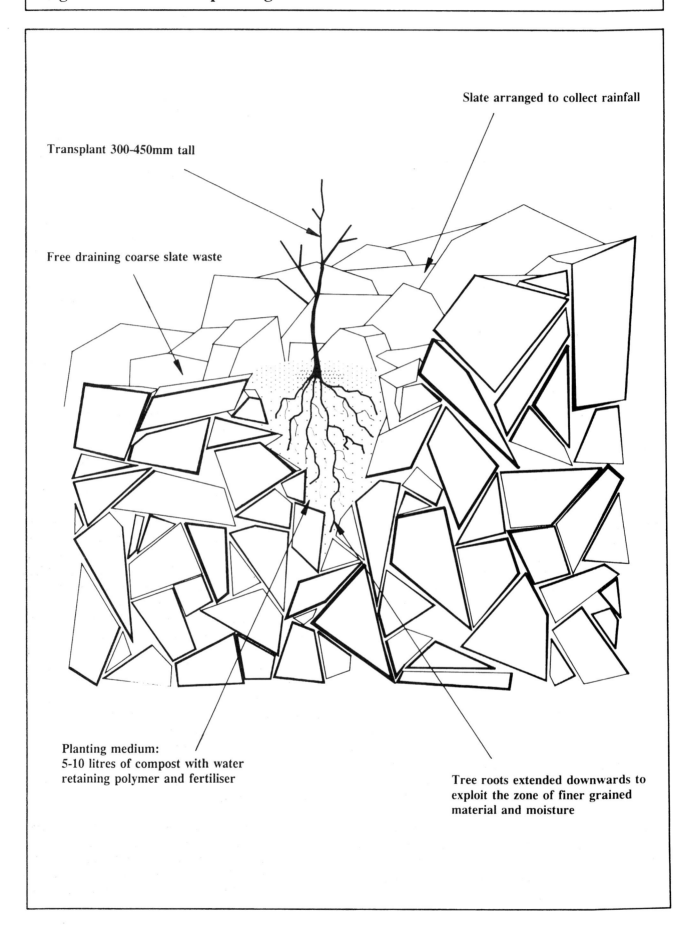

Slate arranged to collect rainfall

Transplant 300-450mm tall

Free draining coarse slate waste

Planting medium:
5-10 litres of compost with water
retaining polymer and fertiliser

Tree roots extended downwards to
exploit the zone of finer grained
material and moisture

tial scheme. The planning and design process should therefore include the identification of the maintenance requirements of the proposed vegetation, so that provision can be made. Where appropriate commitment cannot be made, the vegetation type, site use or even the scheme objectives should be reconsidered.

The management of new grass swards on reclaimed slate waste need not be expensive, particularly in terms of materials. Organic nutrient sources such as sewage sludge and surplus farm slurry are particularly suitable as the nutrient content is released for uptake by plants more slowly and over a much longer period than inorganic fertilisers. Sward management which ensures the continued growth of legumes such as clover will ensure a continued nitrogen supply. Grazing should be managed so that excessive grass growth is recycled as animal droppings and does not become locked up in dead plant material which will be slow to decompose. The intensity of grazing must be balanced with the growth of the grass. It is important to ensure that grazing does not strip away the regenerating parts of grass and clover plants.

Vegetation management also involves the promotion of woodland and other habitats by the timely execution of fencing repairs, fertilising, woodland thinning and similar operations. These are not specific to slate waste or reclaimed land, but appear to have been neglected in the management of most sites.

5.3   Today's approach to reclamation

5.3.1   Evolution of the approach to design

The majority of the schemes outlined in section 5.1 were identified and accepted in principle in the early 1970s but were, at that time, of relatively low priority. The design work was carried out progressively through the 1970s and the early 1980s, before much of the experience described in section 5.1 became available. With the benefit of that experience, and with a knowledge of the current attitudes towards conservation and industrial archaeology it is possible to describe the approach to reclamation which has evolved since the 1970s.

5.3.2   Scheme objectives

Section 5.1 showed the objectives of the reclamation schemes undertaken to date. In most cases they were specific to problems presented by the individual sites, and were developed primarily to tackle existing problems. The schemes could be described as 'problem-led'. The organisations interviewed during this review agree that the most pressing problems have now been tackled. There is therefore a move towards 'opportunity-led' or 'demand-led' schemes in which sites are selected and adapted to provide new facilities or to accommodate new uses. The Welsh Development Agency has adopted a 'strategic' approach to new reclamation projects, in which strategic studies identify demands and opportunities as the first stage in formulating local programmes for land reclamation (Lawday R. pc). One example of this is the recent study of the southern Nantlle valley, covering a complex of some 17 slate workings.

The long-term use of a site, whether 'active' or 'passive', will determine the principal objective of a 'demand-led' scheme. This use should therefore be defined and agreed by all parties involved in the design, funding, project management and long-term management of the scheme and site. Only if this is the case will the objectives of the scheme be properly targeted towards the ultimate site use. Without clearly stated objectives, the design team may be unable to reconcile conflicting interests such as development and conservation, or to determine the appropriate extent of treatments to deal with specific site characteristics. Equally, unless the long-term management and care of the site is determined at the outset the mechanisms and funding provisions may not be allocated or available when needed. Thorough scheme planning by all parties concerned is a pre-requisite of cost-effective site reclamation and reuse.

5.3.3   Changing attitudes

There has been a gradual change in public and professional attitudes towards the conservation of the features of abandoned mineral workings and industrial sites. Department of the Environment advice on the operation of the Derelict Land Grant Scheme (DOE 1992b) states that "the sympathetic treatment of flora and fauna, and features of historic, archaeological or geological interest will also be taken into account. Applicants will be expected to identify and make appropriate provision for such aspects in their reclamation plans". The Welsh Development Agency also expect such consideration to be an integral part of schemes submitted to them although some reliance is placed on planning procedures to assess whether schemes are appropriate in this respect (Griffiths DG. pc). This consideration has been illustrated in reclamation schemes carried out recently combining safety and pollution control with the conservation of mining history at former lead-zinc mines at Mincra, Clwyd and Llanrwst, Gwynedd as part of the Welsh Development Agency's land reclamation programme (Vernon 1989) and in conservation projects at other sites in the care of National Parks, County Councils and voluntary trusts. A similar investigation, consideration and conservation of industrial remains, at a level appropriate to the historical importance of the site and its intended future use, would be carried out in future reclamation schemes for slate workings, assisted by the growing body of specialist literature and expert knowledge.

As with all work involving the rehabilitation of derelict land, there has been a tendency recently towards the use of larger design teams covering a wider range of disciplines than was the case in the 1970s and early 1980s. The initial site investigation, research and feasibility study stages of recent projects have included expertise in many disciplines. For example, the study of the Dorothea Quarry complex, Nantlle was conducted by three firms covering landscape architecture and planning, engineering, and heritage and tourism, and was supported by two specialist reports into industrial archaeology. A further study dealt with industrial waste tipping and hydrogeology.

An abandoned slate working might contain elements of interest to specialists, enthusiasts and the general public, in the fields of:

- surface and underground industrial archaeology;
- landscape, vegetation and wildlife;
- industrial and domestic vernacular architecture;
- geology, mineralogy and geomorphology;
- social history;
- recreation and amenity uses.

Studies such as those at Dorothea Quarry and the Horseshoe Pass, Clwyd have sought to reconcile potential conflicts between the treatment of physical hazards, the removal of eyesores, and the conservation of these features of interest. In both locations there is evidence that public opinion places sufficient value on the conservation of these features to outweigh any perception of an eyesore (Land Use Consultants 1990; Williams 1992). Issues of public opinion and perception are discussed in section 2.13.

5.3.4   The use of new techniques

The WDA-funded research project described in section 5.2.6 identified and developed new techniques for the revegetation of derelict land substrates, particularly in situations where large-scale regrading was not necessary. The techniques make use of native species and vegetation types which mimic those that have developed naturally on some derelict sites, including slate workings. The techniques form part of a 'low-cost' approach to reclamation which may be summarised as seeking to match the existing site characteristics with the requirements of possible new land uses so that major works are minimised, and then selecting vegetation types which will tolerate the existing characteristics or require limited amendment of the site. This more 'adaptive' approach to reclamation has given greater scope for schemes which conserve the existing vegetation and features of interest and concentrate works on essential items of public safety and provision for site use. Reclamation schemes in which the whole area of the site was regraded or disturbed relied entirely on the provision of a new landform, new vegetation and new landscape features such as fences, hedges and walls to create a natural appearance. These large schemes are unlikely to blend fully into the surrounding landscape. Whilst it was appreciated at the time that much greater expenditure on dry stone walls, treatment boundaries, surface undulation and land management would have been required to match the adjoining field and landscape pattern the resources were simply not available. It is recog-

nised that landscape maturity cannot be bought: it requires time. Where a site has mature vegetation or other desirable features, their conservation during a scheme adds greatly to the apparent maturity of the site. Such conservation may complicate the works and add to the scheme cost, in which case it is necessary to balance the value of existing features against the overall financial implication for the project. In other cases the conservation of vegetation or structures can reduce the extent of works necessary to produce an acceptable finished scheme.

5.3.5   Restoration of active slate workings

In Table 2.3 it was shown that over 1000 ha of land in Britain has been worked or tipped on for the production of slate, and a further 544 ha currently has valid planning permission. Little of this land is covered by even basic conditions for restoration. All the quarry operators interviewed expressed their expectation of a continued future for quarrying at their sites. The restoration or rehabilitation of this land is therefore in the control of slate producers, influenced by the current review of IDO permissions and any future 'consolidating' applications or reform of mineral planning permissions granted since 1 July 1948.

5.3.6   The outlook for derelict land schemes

Land no longer covered by planning permission for slate working, and classed as 'derelict' for the purposes of the 1988 Derelict Land Surveys, was shown in section 2.2 to amount to about 600 ha. The reclamation of this land reached a peak of activity in the early 1980s, but has been largely confined to Wales. In early 1994, there were no reclamation or rehabilitation schemes planned for slate workings in Scotland, Cumbria or Cornwall. There were proposals for a number of large sites in Wales, following feasibility studies, but design work had not been carried out. The cost of the reclamation approaches adopted so far, shown in Table 5.1, is one factor in the reduced rate of reclamation activity.

It appears that continuing change in the perception of abandoned slate workings, and that the balance between positive aspects and negative aspects is such that wholesale reclamation is not necessarily regarded as essential or even desirable. Public perception may be cyclical, but no indication has been found that the increased acceptance, understanding and interest in 'heritage' and cultural landscapes has reached its limit. Representatives of the local authorities in which derelict slate workings are found, and grant aiding bodies, were consulted to gain the current views of these organisations on the need for further reclamation and the likely priorities.

It is concluded from these consultations that the objectives of future reclamation and rehabilitation of slate workings will allow a more selective and sensitive approach than the larger-scale approach required to treat hazardous or unacceptable sites in the past. This more selective approach is also better suited to the growing awareness of the wide range of interests which derelict slate workings can support.

The slate working industry and the public authorities which will be responsible for the treatment of derelict slate workings are likely to seek the most economical acceptable solutions, which will often be afteruses such as nature conservation. Where there are no overriding issues of safety, a selective, sensitive approach is also likely to be the most economical.

5.3.7 Reclamation in the future

Future reclamation schemes will, it is believed, adopt a demand-led, selective or adaptive approach which will seek to:

- set objectives;
- identify those site factors which inhibit those objectives;
- identify those site factors which support those objectives;
- implement the minimum works required to overcome the inhibitions;
- conserve and develop the factors which support those objectives;
- manage and maintain the site to sustain those objectives.

5.4 The planning process applied to reclamation schemes

5.4.1 The selection of sites for treatment

In order to monitor the situation regarding derelict land the DOE has required surveys to be carried out at regular intervals, usually every 4 to 6 years. Some periodic surveys have also been carried out in Wales and Scotland. Thus all planning authorities should have access to details of derelict land within their area. Although derelict land has been defined for the purpose of the survey, its selection is still partly subjective and depends on the opinion of an individual.

For example, some slate waste tips in the Lake District are well colonised by vegetation and, in the opinion of Officers of the National Park Authority, would not be improved by reclamation. The definition of "derelict" specifically excludes naturally revegetated areas. Disturbance of the slate waste would remove the vegetation and expose fresh slate waste which would be slow to colonise.

Sites normally selected for treatment have been those which presented a danger to the public, dominated a settlement, marred views from a tourist route or business park or those where a specific beneficial afteruse had been proposed. In the past preference has been given to 'hard' afteruses

such as industry but now equal emphasis is given to 'green' afteruses where significant visual improvement can be expected.

5.4.2 A strategy for a slate working area

Section 6.9 describes some benefits of preparing a strategy, including the ability to respond constructively to opportunities for reshaping tips, using soils or other materials which become available, and identifying future needs for resources. A strategy also helps a local authority to respond quickly to make use of finance which may become available at short notice and for a limited period.

5.4.3 Planning permission

Land reclamation schemes which do not constitute part of the restoration of a permitted mineral working will require planning permission in their own right, and will be considered in the same way as other developments, as described in section 2.4. The introduction of a new land use to a former mineral working will also require a planning permission.

5.5 Financial assistance for land reclamation and project development

5.5.1 Land reclamation assistance

The principal sources of financial assistance for the reclamation of derelict land have been:
- derelict land grant (DLG) in England, from the DOE;
- land reclamation grant (LRG) in Wales from the WDA;
- derelict land grants in Scotland from Scottish Enterprise.

These grants are not payable for works which are the subject of enforceable planning permission conditions or other arrangements providing for restoration. Details of the scope and administration of current grant schemes may be obtained from the organisation listed in Appendix 4.

5.5.2 Priorities for land reclamation

The priority objectives for these grant-aid programmes in England, Wales and Scotland are broadly similar.

England

Derelict Land Grant Advice Note 1 states "Priority within the DLG programme will be given to the treatment of land which in its present condition reduces the attractiveness of an area as a place in which to live, work, or invest, or because of contamination or other reasons is a threat to public health and safety or the natural environment. Such land should be capable of being used to provide for development, amenity value for the community or to contribute towards nature or historic conservation" (para 16). "In rural areas emphasis should be placed on reclamation in areas of particularly high scenic quality or nature conservation value and on schemes to foster development" (para 17(c)). Within this policy framework, DOE regional offices have been able to decide their own priorities.

'English Partnerships' (the Urban Regeneration Agency) took responsibility for administering the Derelict Land Grant and the City Grant as a new unified grant in England from 1994 (DOE 1992c). The proposals gave little indication of intended work in rural areas and so until further details are announced the implications for areas containing slate waste dereliction are unclear.

Wales

The Welsh Development Agency have used a priority list for land reclamation schemes over many years, and have made this widely available (Lawday R. pc). The priorities are:
- schemes for public safety;
- schemes for development end uses;
- schemes for amenity end uses.

Within these priorities, scheme proposals are also assessed on the impact of the existing site on surrounding populations. Thus sites in urban areas or adjoining residential land will be considered more urgent than similar sites in remote areas. In recent years the WDA has increased its emphasis on strategic approaches to 'corridors' and 'gateways' to help guide resources towards coordinated action. The Agency wish to promote a lower-cost approach to reclamation where there are no over-riding reasons for large scale civil engineering works. This approach is set out in the WDA sponsored publication 'Working with Nature' (Robinson Jones Partnership 1987). In 1993 the reclamation programme contained one major reclamation scheme at a slate working, but a number of feasibility studies or strategy studies had recently been commenced or completed in partnership with local authorities and other bodies. These are likely to guide the strategic application of land reclamation and environmental improvement funds to projects which enhance commercial or recreational opportunities and improve the surroundings of residential and business areas.

Scotland

The administration of grant aid in Scotland has undergone a series of changes since the time of the Ballachulish scheme in 1978. The former Scottish Development Agency became Scottish Enterprise, which has now largely devolved its land reclamation function to 13 local enterprise companies. Their priorities are primarily concerned with economic regeneration and employment and so schemes for the reclamation or rehabilitation of slate workings, which are in sparsely populated areas, are

unlikely to be high priorities. Financial contributions are made towards environmental improvements and towards the treatment of derelict land, but the amount of contribution is set at the minimum necessary to make a project viable.

5.5.3 Grants for environmental improvement

The Welsh Development Agency also operate an 'Environmental Improvement' grant scheme, details of which are contained in 'Grants for the Improvement of the Environment -Explanatory Memorandum' (WDA August 1989). This is applicable to, interalia:
- works to enhance the public access to and enjoyment of the countryside;
- works to the outside of sound redundant buildings, for new use;
- the preservation, conservation or restoration of sites or buildings of heritage or archaeological significance, including landscape improvements to their surroundings, and the enhancement of public access to such sites.

Under this grant scheme, higher priority may be given to schemes which fulfil either of the following criteria:
- schemes to provide or improve community facilities on reclaimed land previously subject to grant aid from either the Welsh Office (pre 1976) or the Welsh Development Agency;
- schemes to provide or improve community facilities on derelict land which may be eligible for land reclamation grant, but which would receive low priority due to its nature and location eg the creation of footpaths/cycleways on disused railway land.

Many derelict slate workings for which modest improvement and conservation works would be beneficial, might therefore qualify for grant aid under this grant scheme even if their priority rating under the land reclamation grant programme was low.

5.5.4 Grants for related works

Financial assistance is also available for works to promote nature conservation, woodlands, recreational use and measures which contribute to the rehabilitation of sites. The various grant programmes are outside the scope of this report, but up to date information is available from the organisations listed in Appendix 4. Community based schemes for the rehabilitation of derelict land, the conservation of features of interest, and the enhancement of wildlife conservation, may also qualify for grants from charitable trusts such as the Prince of Wales' Committee and the Shell 'Better Britain' campaign.

5.5.5 Financial assistance for project development

Projects which use slate waste, or use slate workings as a location for new business activity, may be eligible for financial assistance programmes set up to encourage new rural businesses. Most slate workings are located in rural areas designated as Assisted Areas, in either the Development Area or Intermediate Area categories. Development projects may therefore attract grant aid through these schemes. Selective employment grant is available for new developments in Blaenau Ffestiniog (Roberts, pc).

Tourism-based projects have been a particular focus for initial support. A number of initiatives mentioned in this report have received assistance from, or through, the English or Welsh Tourist Boards:

- tourism project initiative at Kirkby in Furness, supported financially by Furness and Cartmel Tourism;
- Horseshoe Pass study, part-funded by WTB;
- 'Industrial and Cultural Heritage' leaflet, produced by South Gwynedd Leader Network, funded by the County Council, North and Mid Wales Tourism, and the European Commission.

5.5.6   Financial assistance for transport facilities

Projects for the extraction, processing and sale of slate waste aggregates have not, to date, received financial support, but grant-aid is available to facilitate the use of rail or inland waterway transport in place of road transport. The Freight Facilities Grant (Department of Transport, 1991) is provided to help firms invest in freight facilities which will reduce the use of environmentally-sensitive roads. These include most single carriageway and urban dual carriageway roads. The grant can assist in the provision of:

- rail sidings;
- wharves and jetties;
- unloading equipment such as cranes and conveyors;
- associated land and buildings;
- rail wagons.

The scheme does not however apply to coastal shipping facilities such as those required for the export of slate-waste from Penrhyn or Dorothea quarries (see Case Study 3).

The amount of grant payable is normally up to 50% of the capital cost of rail or waterway freight equipment, but this may be constrained by the degree of environmental benefit and by the amount needed to make rail or waterway transport more economical than road transport. Transport projects which are commercially justifiable without grant will not attract grant. The appraisal period is normally 10 years and so short-term operations, such as the removal of small slate waste tips, are not usually considered for this grant.

5.5.7   The development of new uses for slate waste

Two further aid schemes seek to encourage novel technology, technology transfer or the implementation of best practice within the environmental field. DEMOS is the Department of Trade and Industry's Environmental Management Options Scheme, which supports collaborative projects which prove the feasibility of adapted or new techniques, and projects which illustrate best practice based on proven techniques. Recycling, and the treatment and disposal of wastes are particular areas of support. ETIS, the Environmental Technology Innovation Scheme, is sponsored by the DOE and DTI. Both aid schemes might be applicable to the development of new uses for slate waste, or new processes to improve waste-derived products.

5.5.8   New arrangements for financial assistance

The grant schemes outlined in section 5.5 are liable to revision or replacement from time to time. Readers are advised to seek up to date information from the organisations listed in Appendix 4.

6 A Framework for the Assessment of the Land Use and Reclamation Potential of Slate Workings

6.1 The purpose of the framework

6.1.1 Objectives

The framework seeks to aid the evaluation of the most suitable land use options for an area of slate working, by identifying the factors and environmental constraints which need to be considered in determining a strategy of treatments for the sites within the area. The framework was developed, tested and refined in the Pilot Study conducted as part of this research project.

6.1.2 Use of the framework

It is intended that this framework will assist:
- local planning authorities responsible for the preparation of local plans, mineral plans and programmes of land reclamation;
- local planning authorities responsible for determining applications for new land uses, further slate extraction or waste tipping or slate waste removal;
- slate quarry and slate waste extraction operators, and others, seeking guidance on the context of their proposals for reclamation and/or new land uses in areas of slate working;
- statutory consultees and non-statutory interested bodies who wish to ensure consideration of their interests at the appropriate stages of the planning process.

The framework provides a step-by-step process to assist these bodies to carry out these functions in a structured manner, with guidance on appropriate information sources, consultations and studies. In particular, it will assist planning authorities in their strategic consideration of areas of slate working. The framework is not intended to prescribe particular uses for slate workings or to rule out other uses. Rather, it is intended to assist in the evaluation and development of a group of sites, so that specific proposals may be considered in relation to their surroundings.

Many elements of the framework can be conducted in outline or in depth, according to the objectives and needs of the user. External factors, specific to the user, can also be introduced into the consideration process.

6.1.3 Underlying principle

The underlying principle of the framework is that an area-wide approach to the management of an area of slate workings will:
- seek to identify the existing values and potentials of sites;
- seek to identify the hazards and other negative characteristics of sites;
- attempt to conserve site value, realise site potential and reduce negative characteristics by the selection of appropriate land uses or treatment for sites;
- evaluate proposed land uses or treatments in the light of adequate site information and a considered land use/treatment plan for the whole area of slate workings.

6.2 The structure of the framework

6.2.1 Stages in the process

The framework consists of a staged process with guidance notes and reference sources to assist the framework user at each stage. These stages, summarised in Figure 6.1, lead to a comparison of the site characteristics found in the study area with the characteristics required by the land uses or treatments under consideration. This comparison identifies both the range of possibilities for each site, and the possible locations for each activity.

The removal of slate waste from disused workings is a temporary land use usually lasting for a few years or exceptionally 10-20 years. On completion the site will require restoration or treatment to facilitate a new use, but the site topography is likely to be radically altered by the removal of material and may be suitable for a wide range of uses which were impossible before disturbance (see section 3.6). The framework therefore incorporates a stage in which the feasibility of slate waste removal is assessed, so that land uses may be considered for the site both in its undisturbed and reshaped states. This parallel consideration is continued throughout the framework.

The site requirements of land uses, and the site characteristics of slate working areas, are recorded in a simplified, standard way so that the two sets of information can be compared. This comparison is facilitated by the preparation of matrices which can be overlaid. The principle is illustrated in Table 6.1. Subsequent matrices, in which increasingly detailed information is shown, are presented as Tables 6.4-6.6.

Figure 6.1 Framework for assessment of land use and reclamation potential

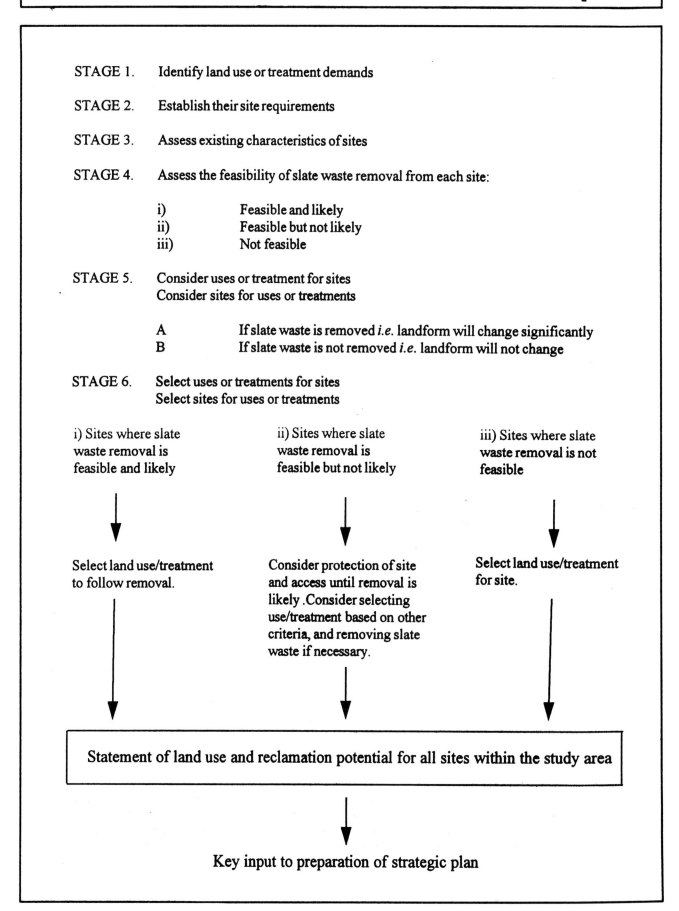

STAGE 1.   Identify land use or treatment demands

STAGE 2.   Establish their site requirements

STAGE 3.   Assess existing characteristics of sites

STAGE 4.   Assess the feasibility of slate waste removal from each site:

        i)   Feasible and likely
        ii)   Feasible but not likely
        iii)   Not feasible

STAGE 5.   Consider uses or treatment for sites
                Consider sites for uses or treatments

        A   If slate waste is removed *i.e.* landform will change significantly
        B   If slate waste is not removed *i.e.* landform will not change

STAGE 6.   Select uses or treatments for sites
                Select sites for uses or treatments

i) Sites where slate waste removal is feasible and likely

ii) Sites where slate waste removal is feasible but not likely

iii) Sites where slate waste removal is not feasible

Select land use/treatment to follow removal.

Consider protection of site and access until removal is likely .Consider selecting use/treatment based on other criteria, and removing slate waste if necessary.

Select land use/treatment for site.

Statement of land use and reclamation potential for all sites within the study area

Key input to preparation of strategic plan

Table 6.1 *Principle of the comparison matrices*

Each characteristic is divided into numbered grades. Each land use and each site is then graded according to requirement or condition.

Matrix 1   Characteristics required for land uses under consideration

	Land Use A	Land Use B	Land Use C
Characteristic 1	1	2	0
Characteristic 2	3	0	1
Characteristic 3	Any	1	2

Matrix 2   Characteristics of slate working sites in study

	Site a	Site b	Site c
Characteristic 1	2	1	1
Characteristic 2	3	3	1
Characteristic 3	1	2	0

The matrices are then compared to see whether land uses and sites match
- It can be seen that Land Use A and Site b are compatible.
- Site a could be appropriate for Land Use A if Characteristic 1 can be modified.
- None of the sites listed appear to be suitable for Land Uses B and C.

6.2.2   The framework

The framework is presented stage by stage in sections 6.3-6.9. Guidance notes are given to identify likely sources of information or to explain the process, with references to section 4 of this report where appropriate.

6.3   Framework Stage 1. Identify land use or treatment demands

Obtain local plans, information on known or perceived land requirements, and information on existing uses of slate workings or proposals for new uses. Where the published information is out of date or insufficient, it may be appropriate for the local planning authority to conduct initial studies or consultations before Stage 1.

Guidance note

Sources include:
- county, district and town/parish/community councils;
- national park authorities;
- development boards, enterprise agencies and similar bodies;
- the Tourist Board;
- operators of existing businesses and ventures;
- landowners.

Compile a list of land uses for which there is, or may be, a demand, and select those which could possibly take place in or near slate workings. Table 6.2 presents some land uses which could be considered. Define each use sufficiently to determine the main requirements in Stage 2.

Table 6.2 *Alternative uses of former slate workings*

Waste disposal	inert material
	domestic refuse
	civic amenity site
Development	industry
	'bad-neighbour' uses
	housing
	craft workshops
	coach and lorry parking
Sports and leisure	outdoor pursuits
	climbing
	mountain centre
	sub-aqua
	motor sports
	shooting
Tourism	caravans/camping
	picnic site
	bunk barn
Forestry	timber
Woodlands	amenity
Wildlife education	conservation
Geology	education
Industrial heritage	education
	tourism
Landscape	
Grazing	

All these uses, described in more detail in section 4, could be considered, but their suitability will be dictated by the degree of site modification required and by external factors such as

6.4 Framework Stage 2. Establish the site requirements of the land uses

Prepare lists of the essential and desirable physical requirements of the selected land uses, with separate columns for:
- the characteristics required of a site in its 'ready to use' condition ie the optimum for the land use;
- the characteristics required of a site from which the 'ready to use' condition might reasonably be produced by site works, investment and development of the facility ie a 'reasonable starting point'.

Classify and grade the requirements in a form that facilitates comparison with site characteristics during Stage 5. The grades selected should represent significant divisions where these influence the feasibility of a proposed land use. For example, interests which are present throughout a site are likely to have a greater influence on site use than individual features of interest which are confined to small parts of a site. Landform components, such as flat areas or voids, which are essential for certain uses should be itemised. It would be possible to be more specific, eg by defining a minimum level area as a criterion, if development was under consideration. The presence of specific types of vegetation, or access to mains water supplies, could also be itemised if needed for particular site uses. The extent of tips could be specified as a volume available for removal (e.g. in order to make processing viable) but determining the volume of slate waste tips accurately is only possible if there is good information about the underlying landform. For the framework assessment, estimates of height and area can be made from published maps and visual inspection. Tables 6.3 and 6.4 give suggested characteristics and divisions as examples, but the list of characteristics and the divisions should be prepared specifically for the land uses selected at Stage 1.

Guidance note

Information sources include:
- published documents and reports;
- trade associations, sport and leisure governing bodies;
- operators and proposers of new land uses;
- feasibility studies and other unpublished documents.

Characteristics can be ranked and listed according to their relative importance or the difficulty of modification. However, the relative importance will vary according to the land uses considered, and the most important factors may be relatively easy to amend during site developments. For these reasons, ranking will not always be appropriate. Ranking could also cause over-complication of the matrices used in Stage 5. If the framework user can apply a degree of technical knowledge to the interpretation of the matrices, the need for ranking can be avoided. Table 6.5 Matrix 1 shows how the site requirements of the land uses should be arranged.

6.5 Framework Stage 3. Assess the existing characteristics of the sites

Identify and distinguish the separate slate working sites within the study area. Sites should be distinguished on the basis of criteria relevant to the selected land uses or treatments, eg mineral planning units, ownership/operation, land areas suitable for the intended uses.

Guidance note

Ordnance Survey maps at 1:10 000 scale, and clear aerial photographs, are particularly useful for this exercise.

Table 6.3 *Characteristics used in framework*

- distance to nearest residences, for consideration of noise disturbance;

- interests eg industrial archaeology, wildlife, geology. Presence may be an asset for, or a constraint on, development. Classification according to degree of interest or constraint;

- adjacent property (buildings, open land, forestry/woodland, water) which may overcome a deficiency within the site or provide a significant additional facility, screening *etc.*;

- landform. The presence or absence of:
  - voids such as pit excavations suitable for waste disposal;
  - flat areas ≤1:10 suitable for development, recreation;
  - steep areas ≥ 1:10 unsuitable for development;
  - undulating topography, *eg* multiple small tips and mixed natural land;
  - enclosure, *eg* shelter from wind, baffling noise, varied habitats;

- access to site, *ie* from within the study area or from existing main routes;

- access within site, *ie* within the site boundary;

- soil or soil-forming material available within or near the site *eg* overburden, subsoil, slate fines;

- vegetation within the site, providing shelter, screening *etc*;

- features/hazards within the site. May be beneficial or a constraint;

- services. The distance to, and availability of, public utility services;

- landscape quality of the setting, influencing site use or enjoyment;

- landscape quality of the site;

- mineral working. Where planning permission exists, its duration and conditions may be major influencing factors;

It may be appropriate to study larger areas in stages, dealing initially with whole sites and then subdividing sites for more detailed study according to the land uses which appear most suitable.

Assess each site to allocate grades for each characteristic selected in Stage 2. This will require a combination of desk study and field work, using existing records and conducting some original work.

Guidance note

Information and guidance may be obtained from many sources, such as:
- published Ordnance Survey maps at 1:10 000 scale;
- aerial photographs, and particularly stereoscopic pairs;
- English Heritage, Countryside Council for Wales, Scottish Natural Heritage;
- RSPB and other non-statutory conservation groups;
- county archaeological officers, county archaeological trusts;

- utility companies;
- county highways departments;
- minerals planning officers.

Where documented information is insufficient or absent it will be necessary to carry out or commission field studies. Guidance on the content and methods of such studies may be obtained from:
- 'Guidelines for the Selection of Biological SSSIs', Nature Conservancy Council 1989 (see also section 4.7);
- 'Conserving our heritage of rocks, fossils and landforms' English Nature 1991 (see also section 4.8);
- landscape assessment method set out in section 4.11 of this report;
- draft 'Guidelines for procedures and standards for archaeological investigation and field work on industrial archaeological sites in Wales' produced by the Welsh Industrial Archaeology Panel, December 1992 (see also section 49).

Table 6.4 *Grading of characteristics*

Characteristic		Grade	Description
Distance to nearest residences		3	< 250m from residence to site
		2	250m - 1000m
		1	> 1000m
Interests- as assets or conflicting with disturbance	*Industrial archaeology; Wildlife; Geology.*	3	Whole site of significant interest
		2	General interest throughout site
		1	Individual features significant interest
		0	No significant interest at the site
Adjacent properties useful for integration to add space or facilities		1	Present
		0	Absent
Landform	*Voids for containment Flat areas < 1:10 Steep areas > 1:10 Undulating topography Enclosed or contained*	1	Present
		0	Absent
Access	*Pedestrian*	3	Excellent (built path or pavement)
		2	Good (unsurfaced path)
		1	Poor (overgrown or undefined)
		0	Working area access restricted
	Vehicular	2	Good (suitable for most vehicles)
		1	Poor (suitable for 4 wheel drive vehicles only)
		0	None
Soil or soil-forming material within or near the site		2	Extensive
		1	Present
		0	Absent
Vegetation within the site		2	Extensive
		1	Present
		0	Absent
Features/hazards within the site		1	Present
		0	Absent
Access to fixed services		2	Present or easily accessible
		1	Difficult to access (500 - 1000m away)
		0	Very difficult to access (>1000m away)
Landscape quality of the setting		4	Very high
		3	High
		2	Medium
		1	Low
		0	Very low
Landscape quality of the site		3	High quality
		2	Average
		1	Poor/damaged
		0	Severely damaged
Mineral working	*Extraction*	2	Active, valid planning permission
		1	Inactive, valid planning permission
		0	None, no planning permission
	Tipping	2	Active, valid planning permission
		1	Inactive, valid planning permission
		0	None, no planning permission
Extent of tips	*Area*	3	>5ha
		2	1 - 5 ha
		1	<1 ha
	Height	3	>15m
		2	5 - 15m
		1	<5m

NOTES ON MULTIPLE GRADES:
Use "/" or "Any" to indicate that alternative gradings are acceptable
Use "+" to indicate that a combination of gradings is essential or present

Table 6.5 *Presentation of characteristics for comparison*

Each characteristic is divided into numbered grades. Each land use and each site is then graded according to requirement or condition. The completed matrices are compared to record the degree of compatibility between sites and use (Table 6.6)

Matrix 1 Characteristics required for land uses under consideration

	Land Use A		Land Use B		Land Use C	
	ready to use	starting point	ready to use	starting point	ready to use	starting point
Characteristic 1	1	1	2	3	0	1
Characteristic 2	3	2	0	0	1	1
Characteristic 3	Any	Any	1/2	3	2	2

NOTE:

1/2, 2/3 etc.	indicates that either grading is acceptable for the land use
Any	indicates that any grading is acceptable for the land use
White land use columns	indicate gradings acceptable in 'ready to use' condition
Shaded land use columns	indicate gradings acceptable in 'reasonable starting point' condition

Matrix 2 Characteristics of slate working sites in study area

	Site a	Site b	Site c
Characteristic 1	2	1	1
Characteristic 2	3	3	1+2
Characteristic 3	1+2	2+3	0

NOTE:

1+2, 2+3 etc.   indicates that more than one grading is present at the site

6.6   Framework Stage 4. Assess the feasibility of slate waste removal from each site

Using the information gathered in Stages 2 and 3, carry out an assessment of the practical feasibility of removing slate waste from each site for use as aggregate, fill material or other bulk uses. Many of the relevant issues are discussed in section 3 of this report.

Classify each site according to the feasibility and the likely timescale of slate waste removal, taking account of medium and longer-term forecasts of secondary aggregate demand and of any known or proposed local demands such as road projects or large-scale land development. The following classes are suggested:

(i)   feasible and likely. Sites which would be considered first by operators and which would be favoured by the mineral planning authority;

(ii)   feasible but not likely. Sites which contain usable material which would be less economic to extract, but would be acceptable to the mineral planning authority. It is assumed that these sites will remain undisturbed until all 'feasible and likely' sites have been worked;

(iii) not feasible. Sites which have little or no usable material and/or where constraints are likely to lead to a refusal of planning permission.

The significance of these classes for the framework process is explained in sections 6.7 and 6.8.

6.7   Framework Stage 5. Consider uses or treatments for sites. Consider sites for uses or treatments

This stage of the framework should be conducted twice: once for sites where slate waste removal is feasible and it

97

may be assumed that the landform will be radically changed, and once for sites which will not be affected by slate waste removal. Sites in class (ii) of stage 4 should be included in both groups.

Where the landform is expected to change radically, the existing landform characteristics should be ignored since the process of slate waste removal could be conducted to leave a new landform designed for the intended subsequent use.

Graded data from stage 2 should be added to the 'characteristics : uses' matrix, and graded data from stage 3 to the 'characteristics : sites' matrix, as shown in Table 6.5. The characteristics of each site can then be compared with the requirements of each land use in turn. This is best achieved by making one copy of the master 'characteristics : uses' matrix for each site, and comparing the relevant column of the 'characteristics : sites' matrix with each land use column in turn. The result of this process is shown in Table 6.6. In the 'ready to use' column of Table 6.6 the comparison of the data indicates:
- ✔ a whole site match;
- ● a partial match where the site has more than one grade;
- x a mis-match.

In the 'reasonable starting point' column, the comparison indicates:
- ✔ a near-match where the site meets this criterion but failed the 'ready to use' criterion;
- ● a partial match where the site has more than one grade and failed the 'ready to use' criterion;
- x a mis-match.

It is not necessary to complete the 'reasonable starting point' column where a match or partial match has been recorded in the 'ready to use' column.

Guidance note

It is recommended that this process is recorded using a system of symbols, such as that shown, to produce a simple visual guide to the difficulty of achieving the selected land use at the site in question. Coloured dots could be used to achieve the same purpose.

The result of this stage will be a sheet for each site, indicating the feasibility of each land use or treatment. The completed comparison sheet for one site in the Pilot Study area is presented as Table 6.7 as an example. To produce a sheet for each land use, indicating potential sites, the converse process should be followed. Make one copy of the master 'characteristics : sites' matrix for each use and compare the data with the relevant column of the 'characteristics : uses' matrix for each land use in turn.

Guidance note

The comparison process can be used in stages to eliminate unsuitable land uses or sites and guide further, more detailed assessment and comparisons.

6.8 Framework Stage 6. Select uses or treatments for sites. Select sites for uses or treatments

Select uses or treatments for sites, from those identified in stage 5. These uses or treatments can be ranked in preferred order according to the likely difficulty and cost of implementation, or other criteria.

Where the objective is to find suitable sites on which to carry out specific land uses, the sites can be listed and ranked according to physical factors, land ownership or other considerations.

Further specific investigations may be required at this stage, or these may be deferred and conducted as part of a subsequent feasibility study or preliminary design.

The selection of land uses should reflect the site classification in stage 4, ie the likelihood of slate waste removal:

(i) sites where removal is feasible and likely in the shorter term:
select land uses or treatments to be implemented after the removal of slate waste. This selection will provide a guide to the restoration/treatment required from the site operator;

(ii) sites where removal is feasible but not likely in the short term:
consider two options:
A protect the site and its access, to permit future slate waste extraction or a later review of the likelihood of extraction.
B select the most desirable land use regardless of landform and, if this requires the removal of slate waste to create a new landform, carry out waste removal as a revenue-generating element of a reclamation scheme.

(iii) sites where removal is not feasible:
select the land uses or treatments which best suit the sites' existing characteristics, to be implemented without the removal of slate waste.

The product of this exercise will be a list of the preferred and/or possible uses/treatments for each site, and a list of sites which are suitable for the land uses under consideration. These lists, together with a supporting statement of the land use and reclamation potential of each site, would form a key contribution to the preparation of a land use strategy for an area of slate workings.

Table 6.6 *Comparison of matrices*

The comparison of Matrix 1 and Matrix 2 is conducted for each site in turn

Matrix 1 Characteristics required for land uses under consideration

Site a	Land Use A	Land Use B	Land Use C
Characteristic 1	[1 X] [X 1]	[2 ✔] [3]	[0 X] [X 1]
Characteristic 2	[3 ✔] [2]	[0 X] [X 0]	[1 X] [X 1]
Characteristic 3	[Any ✔] [Any]	[1/2 ✔] [3]	[2 •] [2]

Matrix 1 Characteristics required for land uses under consideration

Site b	Land Use A	Land Use B	Land Use C
Characteristic 1	[1 ✔] [1]	[2 X] [X 3]	[0 X] [X 1]
Characteristic 2	[3 ✔] [2]	[0 X] [X 0]	[1 X] [X 1]
Characteristic 3	[Any ✔] [Any]	[1/2 •] [3]	[2 •] [2]

Matrix 1 Characteristics required for land uses under consideration

Site c	Land Use A	Land Use B	Land Use C
Characteristic 1	[1 ✔] [1]	[2 X] [X 3]	[0 X] [• 1]
Characteristic 2	[3 X] [X 2]	[0 X] [X 0]	[1 •] [1]
Characteristic 3	[Any ✔] [Any]	[1/2 X] [X 3]	[2 •] [2]

KEY

[✔]	Whole site meets 'ready to use' criterion
[X ✔]	Whole site meets 'reasonable starting point' only
[X X]	Whole site fails to meet either criterion
[•]	Part of site meets 'ready to use' criterion
[X •]	Part of site meets 'reasonable starting point' only

Table 6.7 *Comparison of matrices for land uses at one site in Pilot Study area*

		Land uses						Site
Characteristics		Preparation of aggregates		Civic amenity skip site		Inert waste disposal site		Example
Distance to nearest residences		X	•	•		•	•	2 + 3
Interests – as assets or conflicting with disturbance	*Industrial archaeology*	✔		✔		✔		0
	Wildlife							D.n.a.
	Geology							
Adjacent properties useful for integration to add space or facilities	*Buildings*	X	X	X	X	X	X	1
	Open land	✔		X	X	✔		1
	Water	X	✔	✔		✔		0
	Screening	X	✔	X	✔	X	✔	0
Landform	*Voids for containment*	✔		✔		✔		1
	Flat areas ≤ 1:10	✔		✔		✔		1
	Steep areas >1:10	✔		X	✔	X	✔	1
	Undulating topography	✔		✔		✔		0
	Enclosed or contained	✔		✔		✔		1
Access to site from within study area	*Pedestrian*	X	X	X	✔	X	X	2
	Vehicular	✔		✔		✔		2
Access within site area	*Pedestrian*	X	X	X	✔	X	X	2
	Vehicular	•	•	•	•	•	•	1 + 2
Vegetation within the site	*Woodland/ forestry*	✔		✔		✔		0
	Grass, bracken etc.	✔		X	✔	✔		1
Features/ hazards within the site	*Flooded quarries*	X	✔	X	✔	X	✔	1
	Buildings/ structures	✔		✔		X	✔	1
	Mine entrances	X	✔	X	✔	X	✔	1
Access to fixed services		✔		✔		✔		1 + 2
Landscape quality of the site								D.n.a.
Mineral working	*Extraction*	✔		X	X	X	X	2
	Tipping	✔		X	✔	✔		2
Extent of tips	*Area*	✔		Not Applicable				2
	Height	•						1 + 2 + 3
Other special requirements								

KEY

✔		Whole site meets 'ready to use' criterion
X	✔	Whole site meets 'reasonable starting point' only
X	X	Whole site fails to meet either criterion
•		Part of site meets 'ready to use' criterion

•	•	Part of site meets 'ready to use' criterion part of site meets 'reasonable starting point' only
X	•	Part of site meets 'reasonable starting point' only
D.n.a		Data not available

It is not intended that an apparent mis-match between a site and an otherwise desirable land use should be used to prevent or deter such a use. The framework will assist the initial assessment of the difficulty or extent of works necessary to achieve that desired use. It may suggest an alternative distribution of land uses or may identify sites where slate waste removal as aggregate, or further planned slate waste tipping, would prepare a more suitable landform for the intended eventual land use.

6.9 The preparation of a strategy for the use and treatment of an area of slate workings

6.9.1 External factors

There are many external factors such as local planning policies, economic and employment considerations and the wishes of neighbouring communities which must also be taken into consideration in the preparation of a strategy. Land which has not been disturbed by slate working should also be considered as part of the land use planning process. The strategy would allocate one or more preferred land uses and treatments to each slate working site, or select sites which would be appropriate for land uses known or anticipated to be in demand.

6.9.2 Uses of the strategy

A strategy prepared following an assessment of all the sites within an area can be used to guide developers, operators and other proposers of new uses for land towards the sites which are best suited to those uses as part of an overall approach to land use. The effect that individual proposals or planning applications might have on the potential of other sites, eg by restricting access or by placing sensitive land uses near a site with longer-term potential for noise-generating uses, can be considered.

A strategy can be used to guide the selection of objectives for site reclamation or conservation works, helping to direct resources towards schemes which will have the greatest benefit within an overall plan, and identifying opportunities for partial site treatment through the process of slate waste removal. The site assessments described in stage 3, and the consideration of groups of sites and/or their relative/combined value, are an important step in preparing strategies for wildlife or geological conservation, landscape improvements and making industrial archaeology accessible to the public.

The framework could also be used to provide an indication of the scope and content of the environmental assessment or study needed as part of the design and planning of a new land use, or the commencement of new quarrying activity.

ANNEXE 1:

Case Studies

1.   Kirkby Quarry, Cumbria

2.   Blaenau Ffestiniog, Gwynedd

3.   Nantlle Valley, Gwynedd

4.   Allt Ddu, Llanberis

5.   Ballachulish, Highland

6.   Hodge Close, Tilberthwaite, Cumbria

7.   Welsh Slate Museum and Padarn Country Park, Llanberis

Colour maps and photographs

Case Study No.1 Kirkby Quarry, Cumbria

Summary

Status: active slate production
Operations: extraction, processing, tipping
Ownership: Burlington Slate Ltd
Area: 75ha worked 193ha total
MPA: Cumbria County Council
Issues: planning permission and nature
             conservation
             landscape impact
             secondary mineral supply

CS1.1 Introduction

Burlington Slate took over the quarrying of slate on a commercial scale at Kirkby in Furness since 1843, producing roofing slates from blue-grey slate deposits. Since 1973 the company has acquired six other quarries in the Lake District and supplies a range of roofing and dimension stone products in blue-grey and green slate. All quarried material is transported to Kirkby in Furness for processing centrally. In total, approximately 3550 tonnes of roofing slates and 1100 tonnes of architectural products are produced annually. Walling stone, derived from rivers' stone and the tips, is sold when demand arises. The quarry site extends to 193 ha, which has valid planning permission granted under the Interim Development Order and a smaller 1952 permission.

CS1.2 Nature conservation

The area worked to date is about 75 ha. The site extends over Kirkby Moor, a band of moorland which lies along a ridge of Wenlock Silurian shales. A Site of Special Scientific Interest covering 780.9 ha was notified in 1991 because of the extensive heather moorland, a habitat restricted on an international basis to northern Europe and a scarce habitat within southern Cumbria. Kirkby Moor is the largest area of this habitat in southern Cumbria, and the only area so notified. The SSSI also contains wet heath, mires, flushes, acidic grassland, bracken, streams and rills which add diversity of habitat to the site as a whole. Although the ornithological interest of the site has not been studied it is known to provide breeding areas for peregrine and raven (English Nature, pc). The IDO permission includes 103ha of the notified SSSI area, which is understood to contain commercial mineral reserves. The SSSI constitutes the largest part of the unworked permission, and so quarrying restrictions might affect the economic structure of the operation, yet any damage to the SSSI land would progressively reduce its value and interest. MPG 9 advises that particular atten-

tion should be given to areas such as SSSIs that are within IDOs, and to whether these can be safeguarded or protected. This consideration includes whether there is other land where extraction would be acceptable; and whether operators can voluntarily offer limits on extraction or working areas where this is the only way of avoiding unacceptable environmental damage. However, MPG 9 also advises that there will be limits to what can be achieved without fundamentally affecting the economic structure of the operation.

CS1.3 Landscape impact

The Kirkby quarry contains slate reserves of great depth, and so any waste tipped within the quarry would sterilise usable material. All waste material from slate processing at Kirkby is tipped over the face of a 1 km long tip on the northwest-facing hillside, directly below the quarry. The tip is seen below the hill skyline except from very close viewpoints, and its profile seen from the east is not particularly unnatural, but the tip is prominent from many viewpoints in the southern part of the Lake District National Park. The primary cause of visual impact is the colour of the waste. Freshly-tipped material varies from light grey to light green according to the source of the slate. At any time there is a substantial patch of fresh waste which contrasts with the older, darker material alongside. Old overburden tips above the quarry have regenerated a thin grass sward despite the exposure. A planned scheme of progressive tipping and restoration, which made use of the soil and soil-forming wastes available on site, could do a great deal to reduce the current visual impact and achieve progressive site restoration. A plan for progressive restoration would be a major benefit of a new planning permission to modern standards.

CS1.4 Secondary minerals

Kirkby Quarry has three materials which have potential values as aggregates, bulk fill or raw materials. The 'consolidating' planning application will include the working of these materials. At the boundary of the slate lies a gritstone overburden. Small amounts of this have been sold as roadstone although at present there is insufficient market demand to warrant production. Weathered slate and shale overlies the usable slate reserve. Stockpiles of this material have been screened to produce roadbase and fill. Current production is only 12000 tonnes/year but much greater quantities could be produced if sales enquiries became firm orders. The quarry managers estimate that 1.5 million tonnes of this material exist within the site. The fine-grained residue from screening could be used as a soil-forming material for revegetation works.

Trials have shown this shaley material to be suitable for brick-making but at present there is insufficient demand for bricks to justify investment in a brickmaking plant. However, a nearby plant will soon exhaust its reserve of shale, and could therefore provide an outlet for the Kirkby material. The waste from slate quarrying and processing would be suitable for construction fill and aggregate uses, but at present the local demand is small. Previously, 500,000 tonnes of 'shed waste', ie offcuts and rejects, were supplied for use as fill in the construction of the Greenodd bypass, and 100,000 tonnes of crushed tip waste and waste quarry rock were supplied as 5in.- 0.75in. aggregate for use as free-draining fill in the construction of a British Gas terminal at Barrow. The waste tips contain some 20 mt of slate waste, which could supply demands for fill or aggregate material. The site does have the advantage of proximity to the active British Rail line at Kirkby in Furness. This was once connected by a 1 km spur to the foot of the incline which still stands at the base of the tip. There is therefore the possibility of constructing a conveyor to load rail wagons so that no material need go by road. Rail could then convey the material to areas of high demand, or to Barrow docks for transport by sea. The infrastructure works for rail transport could be eligible for grant assistance as described in section 5.5.

CS1.5 Industrial tourism

The slate-working industry of Cumbria does not, at present, feature in the region's tourist attractions, and the sites of greatest historical interest are relatively remote. Kirkby Quarry could offer visitors a view of the modern slate industry, with historical displays and information about the steps now being taken to protect the environ-ment, conserve primary aggregates and the like. Furness and Cartmel Tourism have expressed their support for such a venture.

CS1.6 Conclusion

The issues of nature conservation, landscape improvement and secondary minerals supply could be brought together, with considerable benefits:
- reduced need for further tipping;
- opportunities to reshape and soften tips economically;
- opportunities to screen wastes, segregating soil-forming materials for use in revegetation;
- enhanced turnover from sales of 'waste' materials;
- substitution for new primary sources of roadstone, fill, aggregates and brick-shale;
- supply at reduced environmental cost.

These benefits would accrue if:
- economic mechanisms made the production of secondary minerals viable;
- a coordinated plan of working, tipping and rehabilitation was approved by the mineral planning authority and implemented by the operator.

The operator is willing to submit a comprehensive planning application which addresses these issues and the possibility of industrial tourism. This willingness is welcomed by the Mineral Planning Authority, as it will enable modern conditions to be introduced to reduce the impacts of working and tipping on nature conservation, the landscape and the environment. It will also secure the restoration of the site.

Case Study No.2 Blaenau Ffestiniog, Gwynedd

Summary

Status: active and disused slate workings
Operations: extraction, processing, tipping
Ownership: multiple
Area: see Figure
MPA: Gwynedd County Council
Issues: landscape and environmental impact
                  tourism and industrial archaeology
                  secondary mineral supply

CS2.1 Introduction

Slate quarrying commenced in Blaenau Ffestiniog in 1760, in what was then a remote mountain ravine. The town developed in the midst of the quarry industry, which remains a major employer. Tourism, linked closely to two working slate producers and the Ffestiniog railway which once exported their products, is another key factor in the town's economy.

CS2.2 Landscape issues

The town of Blaenau Ffestiniog lies in a bowl formed by a ring of hills, extensively quarried and tipped upon. Much of the skyline to the west, north and east of the town consists of slate waste tips which extend right up to the garden walls of some houses. The extreme exposure at this elevation, up to 600m AOD, is a significant constraint on the natural revegetation of the waste tips and workings. The town lies at the centre of Snowdonia National Park but was specifically excluded when the boundaries were set in 1951, as the town and surrounding quarrying were considered to be incompatible with the park. Modern quarrying continues under three companies at a number of sites and although much of this activity is hidden from the town by the topography, some waste tipping on skyline tips continues. Tipping on undisturbed land west of the A470(T) is extending the area affected by quarry working. All slate working in the area is carried out on the basis of post-1948 planning permissions, most of which are old and contain very limited conditions to control working or tipping. Some consolidation of old permissions into more modern permissions has been achieved by the planning authority, and further opportunities are awaited. All visitors approaching the town from the north pass through the main tipping area before reaching the town.

The visual effect of the tips varies greatly according to the weather. Low cloud often obscures much of the surroundings, or dulls colours so that the town and surroundings assume a uniform dark grey colour. In sunlight, the varied colours of hillside vegetation, natural rock bluffs and screes, and slate waste are better defined and so form a mosaic. When the tip outlines are visible the geometric shapes and uniform faces appear unnatural, but in other respects (steep slopes, coarse particle size, lack of vegetation) the tips are similar to parts of the surrounding mountainsides, screes and rock bluffs.

The position of the quarries, some 150-200m above the town, directs noise from blasting and tipping over the town. The dull blast noise is heard daily, but other quarry noise is usually drowned by traffic in the town. If further waste removal was proposed, the generation of noise during quiet evening periods would require regulation. The production of slate granules at one site is subject to tight control over sound emissions, particularly at night time.

CS2.3 Tourism and industrial archaeology

The slate caverns of Llechwedd attract 250,000 visitors each year, and further unknown numbers visit Gloddfa Ganol slate quarry. A significant proportion of these visitors travel on the steam hauled, narrow gauge Ffestiniog railway from Porthmadog. Efforts by the District Council, Mid-Wales Development and Wales Tourist Board to encourage these visitors to explore more of the town are continuing, as they represent a significant source of economic activity in a town with a relatively high unemployment rate and few large employers.

Evidence of the slate industry's history is prominent. From the town centre, the inclines and winding houses which connect the production sheds with former railway sidings are a striking feature which the planning authority has sought to protect. Many of the less accessible remains of older workings are now threatened, or have been obliterated, by modern working. The practice of 'untopping' or converting former underground mines to open quarries, has been a significant cause of losses.

It is suggested by the Friends of the Lake District that continued quarrying threatens the tourist industry in the Lakes area. In Blaenau Ffestiniog the quarrying industry maintains and operates the major attractions to the town. The surrounding mountains have been characterised by slate workings for two centuries. Continued quarrying or mining of slate at those sites and on a similar scale could, with care, be absorbed by the landscape with no impact on further tourism. Uncontrolled opencast extraction and tipping could, however, intensify the problems of noise, dust and visual intrusion to a level which would deter visitors to the town and walkers who enjoy the mountains.

CS2.4 Secondary aggregate supplies

The active slate producers of Blaenau Ffestiniog currently tip some 750,000 tonnes of slate waste each year, of which only a few thousand tonnes is taken for granule production or bulk-fill. Reserves are estimated at over 100 million tonnes, although much of this is located high on the mountains. Large scale slate waste removal and transport by road might cause significant disturbance to the town, and considerable traffic problems. British Rail operate a passenger service to the town, and operate nuclear fuel movements from the recently closed Trawsfynydd nuclear power station. This line could offer an acceptable means to transport all current waste arisings, and some waste from tips, to markets elsewhere in Britain. The long term future of this rail link is in some doubt following the closure of the power station. The rail transport option may therefore have a deadline for commencement, after which the link will be lost. The reclamation of the Glan y Don tip site in 1975 left an area of public open space between the British Rail track and the disused slate loading area (see plan). Part of this area has been developed but space remains which could provide the slate loading facility needed for aggregate shipment. Sufficient space exists for extensive planting to screen the facility from the nearest houses. Material processing facilities such as screens and crushers could be located within the extensive working complexes to minimise noise and visual intrusion.

CS2.5 Land reclamation

The land reclamation schemes at Glan y Don and Fotty tip have removed or reduced the tips which had the greatest impact on residents, due to their proximity to houses. The mass of remaining tips could not, practicably, be reduced substantially unless the exercise was self-funding through secondary material sales, and was undertaken over a period of many years. Reclamation and revegetation are greatly hindered by the sheep which roam freely through the town, quarries and surrounding mountainsides. Fencing quickly suffers damage from vandalism and deliberate cutting, and so grass areas established for public amenity rapidly suffered from overgrazing, even before the young grass took hold. Contractual difficulties caused a lack of early maintenance on the Glan y Don site from which the grass did not recover, and so moss now forms the primary cover. More recent trials of revegetation methods for regraded waste tips, funded by the Welsh Development Agency, showed that deep cultivation to remove compaction, coupled with organic additives such as sewage sludge cake or pulverised domestic refuse which provide a steady release of nutrients, would produce a healthy growth of grass if sheep were excluded.

Case Study No.3 Nantlle Valley, Gwynedd

Summary

Status: active and disused slate workings
Operations: extraction, processing,
                tipping, waste extraction
Ownership: multiple, predominantly private
Area: see Figure
MPA: Gwynedd County Council
Issues: secondary mineral use
                waste disposal
                reclamation
                strategic approach

CS3.1 Introduction

Nantlle lies on the north western edge of Snowdonia National Park. A band of workable slate runs from the north-east to south-west across the valley, between the villages of Nantlle and Talysarn, as shown in the Figure. This slate was worked extensively in deep pit quarries the largest of which, Dorothea, closed in 1970. Smaller scale working continues, and considerable interest has been shown in slate waste removal, waste disposal, and leisure uses of the derelict sites. The reclaimed areas near Talysarn are owned by the Borough Council, but the remaining workings remain in various private ownerships.

CS3.2 Secondary mineral use

The estimated waste stockpile in the Nantlle valley is about 20 million tonnes, of which about half is within the Dorothea complex. Waste slate is currently taken from the Fron Heulog site to supply the Redland Aggregates granule production unit at Blaenau Ffestiniog, since green slate is not available in Blaenau Ffestiniog. It is estimated that 250,000 tonnes of suitable material exists at Fron Heulog, but the rate of extraction is restricted by the planning permission to 15,000t/yr to limit the effects of traffic disturbance, since the only access to the site passes a housing estate. Redland Aggregates wish to extract waste from a nearby site under the GDO but have been refused permission to construct a temporary bailey bridge access from the B4418 across the Afon Llyfni. The company now considers that it may be able to undertake the work under part 19B of the GDO.

The nearby Dorothea site was recently the subject of a proposal to remove 6 million tonnes of slate waste over 20 years. This proposal is described in Box 3.1 (section 3.3.4). The applicant also stated the intention of applying separately for permission to fill two or more quarry holes with controlled wastes. These proposals had generated considerable local opposition on the grounds of

the amenity value, historic and wildlife interest in the existing site, as well as opposition to traffic and landfilling. Objections were raised by statutory consultees including the NRA and Cadw. The proposal was contrary to structure and local plan policies, and was refused planning permission in January 1993 following an Article 14 Direction, not to grant planning permission, from the Secretary of State for Wales.

It is interesting to compare this proposal with the Arup study of Penrhyn quarry's aggregate supply position (Arup Economics and Planning, 1991b) which concluded that a 1 mtpa scheme was marginally non-viable even if the capital outlay was excluded ie covered by grant. On this basis the shipping of 200,000 tpa from Dorothea via Port Penrhyn, Bangor would appear to be less competitive due to the additional 10 miles (16 km) road haulage costing approximately 50p/tonne. If the harbour infrastructure costs are included, the Dorothea scheme appears significantly less competitive, indicating that the separate refuse disposal application would have been crucial to the economic viability of the proposal.

This comparison, however crude, illustrates the difficulty of achieving a viable scheme for the long distance shipment of slate waste to areas of demand. In Arup's estimates, the ex-works costs of slate aggregate were no more than 10% of the eventual 'ex-wharf' cost in the Thames estuary. Shipping was the largest single operating cost, but road haulage from quarry to port formed over 15% of the operating total. On this basis, large-scale, long-distance shipment of slate waste can only be economically feasible where slate waste reserves and transport infrastructure exist in close proximity.

A smaller-scale proposal to remove 25,000 tonnes per year from the Ty Mawr East site (see Figure) has been approved. The material would be used predominantly by the applicant, a local building firm. The extraction was planned to minimise the transmission of noise from the site, and to conserve or record historic remains in conjunction with the County Archaeological Trust. The developer also proposed the disposal of inert wastes, which was granted a separate planning permission.

CS3.3 Waste disposal

The issue of waste disposal in disused slate quarries is particularly focused on Nantlle valley, since within a 1km radius there are examples of:

- an active refuse disposal site;
- a flooded quarry containing industrial wastes;

- controversial proposals for large scale refuse disposal;
- planning permission for inert waste disposal.

The waste disposal proposals are both linked to slate waste removal schemes described in CS 3.2. Under the two-tier system of local government, the county councils in Wales are the mineral planning authorities but waste disposal requires planning permission from the borough or district council, and so separate planning applications are required. These applications are then determined separately.

The issues raised by waste disposal activity and proposals include:

1. A lack of information about the dispersal of leachates from unlined quarries. It is understood that investigations have commenced, but no confirmation has been obtained. The NRA require that the remaining Cilgwyn quarry be lined before landfilling, and would apply the same standards to other quarries at Nantlle.

2. Proximity to houses – see Figure.

3. Landscape impact. The Cilgwyn waste disposal site is largely shielded from view by the topography and waste tips. The Dorothea site, by contrast, lies in the valley floor and is overlooked from many houses. The removal of slate waste would reduce the low-level screening presently existing.

4. Industrial archaeology. Landfilling of these quarries would reduce the integrity of the setting of important remains, as explained in the next section. The Cornish Beam Engine House, a Scheduled Ancient Monument, would lose much of its relevance if the quarry hole it once kept dry by pumping was to be filled and covered over. Similarly, the 'pyramids' and other transport system remains would cease to relate to the site.

5. Road traffic. All waste brought to the site for disposal would be imported by road, with associated effects of noise, traffic disruption and road safety. While it would be possible to transport slate waste and domestic refuse using the same vehicles and ships, particular care would be needed to avoid cross-contamination and to ensure that refuse loads were sealed. The transport of inert waste, eg rubble, would not cause cross-contamination problems.

6. Postponement of existing reclamation plans for Dorothea, for the duration of the landfilling and capping phase. The current reclamation proposals are for the creation of a country park based on the quarrying history of the site. Landfilling would significantly affect the fundamental character of the site and thus its subsequent use.

CS3.4 Reclamation

Two reclamation schemes were carried out near Talysarn, at the western end of the slate-working complex: Coed Madog in 1972 and Talysarn 1978-79. These schemes were then seen as the first phase of a progressive clearance programme which would extend through Dorothea towards Nantlle. The current reclamation proposals for Dorothea would continue that process although possibly in a less radical manner. The continuing development of public and official attitudes towards industrial archaeology, 'cultural landscapes' and sensitive approaches to reclamation, described in section 5.3, suggests that should reclamation proceed, the 1988 feasibility study for Dorothea (Lovejoy and Partners 1988) may be reviewed and refined to take account of changed attitudes. Certainly the detailed development of any further reclamation would be expected to draw on the newer revegetation techniques now available, which would permit much of the site's vegetation and historical interest to be retained. The need for reclamation to include an extended aftercare programme to ensure the proper development of new vegetation, and for close attention to design details which respect and relate to the surrounding landscape, is now clearly recognised. The capital cost of the 1988 "quarry park" proposals was estimated at £6.75m of which £3.22m would be in the form of land reclamation grant (Lovejoy and Partners 1988). The feasibility study anticipated an annual operating deficit of over £100,000. The acceptability of such a deficit is questionable when the areas already reclaimed are gradually deteriorating due to a lack of aftercare. If the modest resources needed to repair fences and arrange an annual fertiliser or sewage sludge dressing are not available to the local authority, it seems unlikely that £100,000 per year can be found to underwrite the forecast deficit. A more modest, sustainable approach is likely to be adopted by the local authority if the site becomes available to them.

CS3.5 Strategic approach

The number and range of proposals, planning applications and activities currently concerning this relatively small but intensively quarried area, together with the range of possibly conflicting issues surrounding each site and activity, make some form of strategic approach crucial to the rational consideration and management of these pressures. The combination and interaction of pressures may have far greater consequences than isolated consideration might suggest. For example, traffic flows on the B4418 are a concern common to many sites and proposals, and although some HGV traffic generation may be acceptable the carrying capacity of the road will be reached at some point. If the mineral planning authority and/or the Borough Council are presented with an application for planning permission, in the absence of information or proposals for surrounding sites, the determination of the planning application in isolation may prejudice more desirable future activity nearby. Equally, full consideration can only be given to the nature conservation, historical and amenity value of the many abandoned slate workings if adequate time is allowed for a valid assessment of the proposal site and its relationship with nearby sites. To date this information is incomplete.

The Planning Department of Gwynedd County Council, and the Welsh Development Agency have sought to address this lack of information by commissioning a study of the part of the Nantlle valley workings lying south of the B4418.

"The purpose of the study is to prepare an appraisal of the derelict sites and recommend outline proposals for:

- making the sites safe for visitors;
- environmental enhancement of the area;
- preservation of worthwhile industrial archaeological features;
- the future use of the area by fauna and flora" (study brief 1992).

The study included surveys of tip stability, hazards to the public, archaeological remains and wildlife habitats, and an assessment of the environmental impact of the sites on the nearby villages. This comprehensive approach to the study and future treatment of slate workings provides a good example of the wider considerations now thought desirable.

Case Study No.4 Allt Ddu, Llanberis, Gwynedd

Summary

Status: disused slate working, reclaimed
Scheme cost: £1,600,000
Ownership: Arfon Borough Council
Area: 24 ha
MPA: Gwynedd County Council
Issues: purposes of reclamation
                  reclamation methods
                  landscape and industrial archaeology

CS4.1 Introduction

The pit and hillside workings of Allt Ddu are known to have begun before 1771, and in 1809 they were absorbed into the massive Dinorwig undertaking although Allt Ddu and the nearby Chwarel Fawr continued to operate as a separate department. A land reclamation scheme was carried out between 1982 and 1988, although much of the design work had been completed by 1976.

CS4.2 Purpose of reclamation

The site consisted of two large, flooded quarry pits, together with several smaller pits, derelict processing and quarry buildings, tramways and tips. The quarries were unfenced, and skirted by the public road serving houses around the site. Buses used the narrow road between two quarry holes, with little room to manoeuvre. The dangers to pedestrians and vehicle occupants were considered unacceptable. The tips and derelict buildings were thought to present an unacceptably degraded environment for the adjacent houses. The site lies on the north-west fringe of the Dinorwig complex and extended the influence of dereliction considerably towards Llyn Padarn. The objectives of reclamation were to remove the hazards presented by the quarry holes, to remove the dereliction of tips and structures, and to improve the road and drainage services to the adjoining properties. As initial proposals for small residential plots did not find support, it was decided to produce woodland (4 ha) and grassland (20 ha), to be managed for amenity use.

Since the completion of the scheme the grassland has been overgrazed by sheep despite the Borough Council's attempts to have them removed. Footpaths across the site are well used, and low intensity informal use of the site continues. The local running club include part of the site in their courses. Rock climbers are frequently to be seen using the remaining exposed face of the Allt Ddu quarry. Access to the foot of the face is easy and so the climb is popular, although for reasons of liability the Borough Council do not officially permit climbing.

CS4.3 Reclamation methods

In essence, the scheme was simple. The quarry holes were pumped dry to allow the removal of dumped cars and similar material, and then slate waste from the tips was returned to the quarry holes. 1,100,000 mt of waste was moved, primarily by excavator and tipper lorry although dozers were used for short movements and for grading. Compaction was achieved with vibratory rollers. The new landform resembled that before quarrying although the tip was not completely excavated to ground level. Slate waste was used to create an improved road alignment where previously it skirted the quarry holes, and crushed slate was screened to produce the Type 1 sub-base used in the road construction.

Topsoil or subsoil were not available as a treeplanting medium, since little soil remained at the site and no major construction projects were in progress in the immediate vicinity. This situation applied to most slate waste reclamation projects. The slate waste surface was therefore crushed using at least 10 passes of towed grid rollers of 15 tonnes weight. Lime and phosphate were applied, a fescue and clover-based seed mixture was broadcast and broiler house litter was applied as an organic fertiliser and surface mulch. This technique was, by that time, standard practice for grass establishment. There was no standard technique for tree planting and since it was intended to establish 4ha of new woodland, soil importation was unrealistic. The solution was to 'manufacture' a raw soil substitute by crushing selected fine slate waste to the following size specification:

mesh aperture	minimum % by mass passing
25mm	100
10mm	95
1mm (medium sand)	50
0.02mm (silt and clay)	20

This material was placed and lightly compacted to 250mm thickness over the planting areas. Transplants of Alder (Alnus incana), Willow (Salix cinerea), Rowan (Sorbus aucuparia), Birch (Betula pendula) and Oak (Quercus petraea) were planted in pits 450mm dia, 300mm deep, backfilled with a 2:1 mixture of the crushed slate and planting compost to which slow-release fertiliser was added. Polythene sheet mulch mats, 1m², were laid around each plant to reduce the evaporation of moisture from the substrate. The initial establishment rate was good although some losses were caused by abrasion of the tree at ground level in strong winds, and by feral goats. The Alder, which are nitrogen-fixing, have continued to grow strongly and will aid the other species by providing shelter and promoting soil development. The bare slate surface between the trees has now been colonised by thousands of self-sown Birch and Willow from nearby woods.

The grass and planted areas have received no maintenance since the completion of the initial contract in 1988. Although the trees would benefit from a further fertiliser application they have not suffered significantly. The grass has been grazed without permission, but no fertiliser has been applied since the initial top dressing. As a result the clover and grass have declined and largely been replaced by moss, as has occurred elsewhere. This again demonstrates that reclamation requires a programme of works, after the initial civil engineering, to ensure gradual, maintained soil development. This is recognised in other mineral operations such as opencast coal or sand and gravel working, where restoration of soils to agriculture follows a five-year aftercare programme. If vegetation is to survive on slate waste, a similar programme of soil development and vegetation management is the minimum requirement. This could include regular applications of organic materials such as sewage sludge, for which expenditure would be modest, and the proper rotation of grazing livestock to ensure sward recovery between grazing periods. There is a need for a forward commitment of modest finance and staff resources at the beginning of the reclamation process, to ensure that reclamation achieves its objectives. Where this commitment cannot be made, examination of the land use objectives may identify an alternative vegetation, such as woodland or scrub, which is less sensitive to a lack of long term care.

CS4.4 Landscape and industrial archaeology

This reclamation scheme was designed at a time when much of today's concern for cultural landscapes and the conservation of industrial archaeology had not developed, and the over-riding need was seen to be the clearance of dereliction. One of the early design ideas for Allt Ddu was to create an intimate landscape of individual low-roofed cottages with rendered walls, slate roofs and dry-stone walled enclosures typical of the nearby scattered community of Fachwen. The necessary earthworks, services, dry-stone walling and tree planting could not be funded at the time, but are still possible and for this reason the scheme should be considered as incomplete. The area of open grassland is substantially larger than the surrounding small fields, and has a relatively bland surface texture unlike the small-scale undulation and variation of the natural grassland. The retention of a bluff of rock, which projects from the underlying ground and was uncovered in the excavation of the tip, does provide some visual interest in the uniformly regraded slope. The vegetation is also too uniform and of an unnaturally pale colour, although this is largely due to lack of management. Even if the introduction of greater topographical variation by minor earthworks is not considered to be justified now, the opportunity still exists to add the stone walls or enclosures and vegetation interest which is currently lacking. The Borough Council, which owns the site, proposes to add a patchwork of woodland and traditional boundaries, as the basis of a more positive management and vegetation development regime, when grant aid is available.

The reclamation scheme was carried out with little or no investigation of the historical interest or significance of the derelict buildings and quarrying structures of the site, and consequently all except one powder magazine were lost. The former mill building, disused since 1870, and an interesting transporter incline, may have been worthy of retention to add to the recreational and educational value of the site. A similar site today would undoubtedly merit a thorough archaeological survey before any large-scale regrading was designed.

CS4.5 Conclusion

This reclamation scheme demonstrates that the techniques exist to make dramatic changes to abandoned workings and their surroundings. The skills of professionals in many disciplines can ensure that such changes are integrated with the surrounding, high-quality, landscapes provided that sufficient resources are made available. Such projects might involve cooperative funding from a number of agencies, with a range of complementary objectives, working over several phases of reclamation and landscape development. The need for continuing care and development of the new landscape and vegetation created is also highlighted.

Case Study No.5 Ballachulish, Highland

Summary

Status: reclaimed slate working
Scheme cost: £1.3 million
Ownership: Highland Regional Council
Area: 25 ha
Issues: impacts and opportunities study
                 reclamation techniques
                 continued management

CS5.1 Introduction

Ballachulish stands on the south shore of Loch Leven near Glencoe, in an area of natural beauty. Slate quarrying began in 1693 (Gilchrist 1981) and continued until 1955. The need to reclaim the dereliction left by this industry was discussed for over ten years but until the Scottish Development Agency was formed in 1975 no body existed with the resources to take action. The reclamation scheme was undertaken in 1978-79, and was recognised by a Civic Trust award in 1980 and a Europa Nostra diploma in 1981.

CS5.2 Impacts and opportunities studies

The scheme was preceded by a feasibility study prepared in 1977, which proposed land reclamation and environmental improvements within a framework for commercial and industrial development set by the SDA. The feasibility study drew on:

1. A carefully defined brief and timetable, including a considered statement by the planning authority of likely future land uses and key sites;

2. Sub-regional and area studies, and the Scottish Tourist Board's programme for tourist development in the village;

3. Detailed suggestions for Ballachulish by the Countryside Commission for Scotland, in the context of their local work;

4. The expectation of a major improvement to the A82(T) road, creating development opportunities;

5. Strong private interest in the provision of hotel and chalet accommodation.

The feasibility study was therefore able to identify and address the impacts of the dereliction with a clear understanding of the land use needs of the site and village.

The principal impact of the dereliction was from the extensive slate waste tips which covered the lochshore and hillside around the quarries, and loomed above the houses of East Laroch. The waste tips obscured views over Loch Leven, and gave the whole village a depressed appearance. The disused buildings, and generally unsightly uses of the former British Rail station yard, added to the unattractiveness of Ballachulish although its location on the A82, a busy tourist route, gave the village considerable potential.

The study proposed a programme of filling and regrading to alleviate the impact of the slate waste and to provide the landforms needed for the proposed developments. The large overall site and the opportunity to tip some additional material into the loch gave a small extra degree of flexibility in the landform design. In all, it was necessary to move about 750,000 tonnes, or one-tenth, of the waste to achieve the scheme objectives.

CS5.3 Reclamation techniques

Representatives of the SDA and their consultants visited schemes carried out in North Wales, and were able to adapt some of the techniques to suit the particular conditions of the site. The engineering works were generally the same as those described for North Wales, except that draglines were used to regrade the steepest waste tips since disturbance at the base would have been de-stabilising. Heavy winter rains made work on these slopes particularly hazardous and two landslips occurred. A significant quantity of waste was used to fill two flooded pits in the quarry floor. These pits, each about 25 metres deep, were retained as shallow pools with the quarry faces as a dramatic backdrop. Slate waste was also used to extend the foreshore to permit development, and to produce the formation for the new A82 (T) route. New foul and surface water drainage systems were constructed.

Revegetation consisted of grass establishment and tree planting. The Ballachulish slate is pyritic, and oxidation of pyrite within the waste had produced acidic conditions. All areas were treated with ground limestone at 5t/ha. Trials of surface preparation methods led to four methods being used in different circumstances:

- imported topsoil, which was restricted to priority locations on cost grounds;
- lochside soil, which was excavated to provide additional waste tipping capacity and was amended with peat, gypsum and fertiliser;
- hydraulic seeding, which although less satisfactory was the only practical means of seeding the steepest slopes;
- broiler house litter, which was economical and effective.

Two grass seed mixtures were used, one selected to cover planted areas with a minimum of competitiveness, and the other to produce an amenity sward tolerant of minimal maintenance and light grazing. Over 80 000 trees were planted, predominantly into beds of topsoil or lochside soil. The species selected were 'pioneers' able to withstand wind exposure and some drought:

Ash	Fraxinus excelsior
Alder	Alnus glutinosa
Birch	Betula pubescens
Larch	Larix europaea
Hawthorn	Crataegus monogyna
Willow	Salix caprea

CS5.4 Success and continued management

The initial success of the scheme is indicated by the awards noted in CS5.1, but the scheme has also stood the test of time. The major hotel development did take place (after some doubts) and the trunk road has been re-aligned, providing the space for a community building, tourist information centre and car parking. The regraded lochshore tips are used for informal recreation, and further harbour or marina facilities could be added. The quarry and adjacent tips remain a distinctive landscape feature but do not seem out of place at the base of the mountainside.

The reclaimed, grassed tip matches the background of open hillside and forestry much more closely than the small-scale enclosure landscape found in parts of Wales (see Case Study 4.4).

The grass areas and planting are monitored by the Regional Council, who have controlled the degree of grazing. As a result there is a balance between the light grazing pressure and the limited growth potential of the sward. Clover has survived, contributing some nitrogen, and sward regression has been avoided with no need for fertiliser additions. Moss has developed in parts of the sward, but this is probably the result of seasonal extremes of wetness and dryness. The sward is open enough to allow heather colonisation and some spread of tree seedlings, which will contribute to a natural appearance and wildlife interest. The presence of Rhododendron ponticum seedlings is cause for concern as this shrub is extremely invasive and persistent. The trees have grown strongly to form young thicket woodland which would benefit from some thinning and species enrichment.

CS5.5 Conclusion

This reclamation scheme shows the benefits of a clear direction and long-term objective for a scheme, in addition to the desire to tackle existing problems.

Case Study No.6 Hodge Close, Tilberthwaite

Summary

Status:	active and abandoned quarries
Operations:	extraction, tipping, processing
Ownership:	multiple private
Area:	Hodge Close 15 ha disturbed, approx Moss Rigg 1.2 ha quarried, 6 ha permission
MPA:	Lake District National Park Authority
Issues:	control of workings unsuitable aggregate sources new recreational uses

CS6.1 Introduction

The small valley of Tilberthwaite lies 5 km north of Coniston, in the heart of the southern Lake District. Numerous small slate quarries and mines are known to have operated, but the largest areas of remains are at Hodge Close and the nearby Moss Rigg (see Figure). Active working continues, but substantial natural colonisation and informal recreational activities have 'reclaimed' parts of Hodge Close.

CS6.2 Control of slate working

Moss Rigg quarry is owned and operated by Burlington Slate Company Ltd, and its most recent planning permission, granted in 1985, extends over a total of 6 ha. The site is largely surrounded by birch/oak woodland which obscures the open quarry and reduces the visual impact of the tips. The site is most clearly seen from the Hodge Close complex. The most prominent sign of recent activity is the modern workshop building on the plateau of the lowest tip. No extraction takes place at present but the site will be worked when demand warrants. All usable slate block would be taken to Kirkby in Furness for processing. The planning permission does not provide for the restoration of the whole site, but does serve to control the location and extent of waste tipping.

The Hodge Close site was abandoned before the current slate extraction commenced. A small-scale operator started extraction from the adjacent Peat Field Quarry, and has reworked material from part of the extensive waste tips, sorting and splitting by hand to supply roofing and paving material. As the waste tips are indistinct the planning authority was unable to determine whether the GDO 2 ha limit applied or not. They decided to control the volume and method of transport by a section 106 agreement which will also limit the scale of working. The processing is carried out within the old buildings of the site and at this scale causes no disturbance.

CS6.3 Unsuitable aggregate source

The extensive waste tips at Hodge Close and Moss Rigg contain substantial quantities of potentially usable aggregate or fill material, but the removal of this material would cause great disturbance. Although there are only a few houses in the vicinity these lie alongside the very narrow, single track road which serves the quarries. This road is already used by many recreational visitors to the site including minibus parties. The valley is free from most sources of noise, although the small scale quarrying and visitor vehicles do generate intermittent noise. Most of the waste tips have not been disturbed for many years and have weathered to a dull grey-green colour, reducing their prominence. Many mature and young birch trees have developed on the finer-grained tip material and patches of undisturbed ground. These trees also soften the appearance of the waste and workings.

At present there is no likelihood of a demand arising which would generate interest in this source of material. This site is, however, a good example of the natural 'reclamation' which is slowly occurring elsewhere and which would be threatened by any large scale disturbance.

CS6.4 New recreational uses

Hodge Close is a centre for unofficial activities such as rock climbing, abseiling and sub-aqua diving. The flat waste tips provide suitable parking for minibuses and cars adjacent to the quarries. The Tilberthwaite valley is also crossed by way-marked footpaths connecting National Trust land and the surrounding fells.

Despite the limitations of the narrow road access, the site is valued because it is relatively robust and takes the pressure off other, more sensitive, locations. 'The National Park Authority would wish to see some form of management of the activities on the site rather than their active promotion or intensification of use. Part of the site is now registered as derelict land and would be eligible for a derelict land grant'.

CS6.5 Conclusion

The effects and impacts of slate workings are sometimes closely related to scale. In this case, continued small-scale working in a relatively traditional manner can be absorbed by the site with no further impact. The scale of the tips, softened by natural colonisation, is such that they do not dominate the landscape. The scale of recreational use is within the "carrying capacity" of the site. It seems that the present situation does not cause significant problems,

and so provided that the scale of activity is regulated there
is no need for action to change the status quo.

Case Study No.7 Welsh Slate Museum and Padarn Country Park, Llanberis

Summary

Status: former slate quarry complex
Ownership: Gwynedd County Council
Issues: collaborative action
              conservation
              new site uses

CS7.1 Introduction

The Dinorwig quarries complex was one of the largest in Britain, producing over 100,000 tonnes of slate per year in the late 1890s. The scale of the workings, which spanned 2.5km and over 600m vertical difference, can be seen in the Figure. When the Dinorwig quarries closed in 1969 the site, its workshops and their contents were auctioned. The former Caernarfonshire County Council and the Ancient Monuments branch of the Welsh Office were aware of the historic importance of the site and particularly the Gilfach Ddu workshops erected in 1870. The County Council bought the part of the site which consists of:

- Vivian quarry;
- Gilfach Ddu workshops;
- Gilfach Ddu station and former loading yards;
- lakeside tips;
- quarry hospital;
- Alltwen oak woodlands.

The major part of the Dinorwig quarries complex was acquired by the CEGB for the construction of the Dinorwig pumped-storage hydroelectric generating station.

CS7.2 Collaborative action

The site now offers a country park, recreational activities, the Welsh Slate Museum and the Vivian Quarry. This was achieved by collaboration between local government and national bodies. Following the purchase, the County Council agreed that the Welsh Office would take responsibility for the care of the fabric of the Gilfach Ddu workshops which are a Scheduled Ancient Monument, the National Museum of Wales would be responsible for the contents of the building, whilst the Council would be responsible for the remainder of the site. CADW later acquired two of the incline series, both Scheduled Ancient Monuments, and have listed other structures for which they are now responsible.

The National Museum had already begun a collection of slate artefacts and was seeking an outpost in which to pre-sent them. They were therefore particularly willing to take part in the collaboration at Gilfach Ddu, to set up the Welsh Slate Museum.

Gwynedd County Council now manage and promote the Padarn lakeside as 'Llyn Padarn Country Park', providing countryside access within easy reach of the visitor to, and residents of, Llanberis. The lakeside tips of Vivian Quarry were partly reshaped in 1972 with funds from the Welsh Office, to improve car parking and public access. This car park serves both the museum and the recreational activities of the Country Park. One conical lakeside tip was originally formed by tipping from a tramway, which passed over the railway on a timber bridge. The tip was reduced in height, and the bridge replaced, to form an accessible viewpoint over the lake.

CS7.3 Conservation

The National Museum of Wales had a policy objective of setting up outposts which represented the primary industries of Wales' industrial development. Slate was one of these industries. The Museum wanted these outposts to be in the key locations and settings of the industries rather than at unconnected sites selected for their accessibility. The opportunity of Gilfach Ddu was doubly valuable since Dinorwig was, with Penrhyn and Blaenau Ffestiniog, a primary slate quarrying location but was also an established tourist centre. Dinorwig was also relatively 'unspoilt' by the most modern working methods, and retained all elements of extraction, waste tipping, transport systems, processing, workshops and rail link to form a valuable educational and interpretive resource.

The museum workshops retain an active use in the conservation and repair of machinery, tools and artefacts for display. Many of the staff are former quarrymen, and their practical demonstrations and first-hand knowledge are much valued by visitors.

The CEGB took care to minimise the damage and disturbance caused by the construction of the power station, including the dismantling and re-erection of a table incline. Recently an unstable retaining wall, protected by listing by CADW, was stabilised in-situ with financial aid from the Welsh Development Agency. The cheaper option of demolition and regrading was rejected.

The wildlife value of the workings and the undisturbed habitats within the complex is enhanced by its management as a country park.

CS7.4 New site uses

In addition to the Museum, and the nearby underground power station, the site offers visitors a range of activities. Vivian Quarry is accessible on foot, and contains a 'blondin' transferred from elsewhere within the site. The quarry hospital and its displays are a popular attraction. Many of the inclines, winding houses, and quarry areas are accessible via waymarked paths. Interpretive leaflets cover the quarrying, wildlife and history of the site. The narrow gauge railway which formerly transported quarry products to Port Dinorwic on the Menai Straits, is now operated as a tourist steam ride along the lake shore. Canoeing and watersports are available at the lakeside, near the car park. Rock climbing is reported to take place in Vivian Quarry, although it is not clear whether this is officially permitted.

The National Grid Company, who succeeded the CEGB as operators of the pumped storage station, include minibus tours to parts of the main Dinorwig quarry complex as part of their visitor tours.

CS7.5 Conclusion

This site is an excellent example of the way in which a number of bodies can work together to make maximum use of the resources offered by an abandoned site in a way which shares the management costs, involves a range of expertise and interests, and interprets a feature that dominates the north-west side of this valley.

Kirkby Quarry, Cumbria

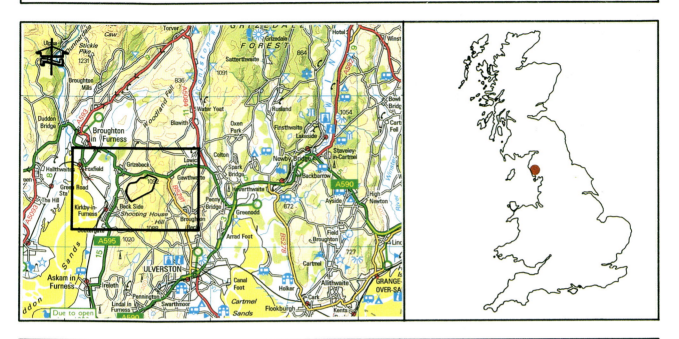

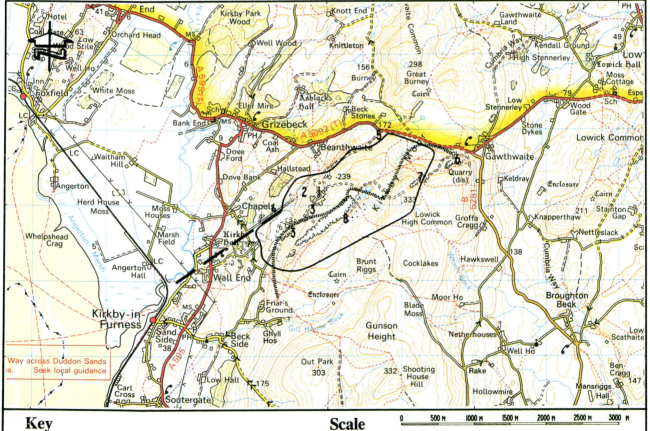

Key

Scale

1. **Active tip now encroaching on footpath**
2. **Existing woodland screen could be enlarged into field**
3. **Disused waste tips**
4. **Incline link to rail system at Kirkby in Furness (disused), route shown thus:** — — — —
5. **Main road access**
6. **Quarry recorded as derelict in 1988 DoE survey**

7. **Site of Special Interest (780.9ha), Boundary shown thus:** ◦◦◦◦◦◦◦◦◦◦◦◦◦◦
8. **Active quarry**

1. Quarrying and processing operations at Kirkby Quarry. Land behind the quarry forms part of an SSSI. The former tramway link to the railway is visible to the right, near the caravan park. Tips of fine residues from settlement lagoons can be seen in the left of the photograph, and waste rock is currently tipped to the right of the processing buildings.

PHOTOGRAPH COURTESY OF BURLINGTON SLATE LIMITED.

2. This waste tip could eventually cover the fields at its base. The largest blocks roll furthest, forming a layer beneath the tip.

3. Modern extraction methods enable slate to be extracted with less waste than conventional blasting techniques.

Blaenau Ffestiniog, Gwynedd

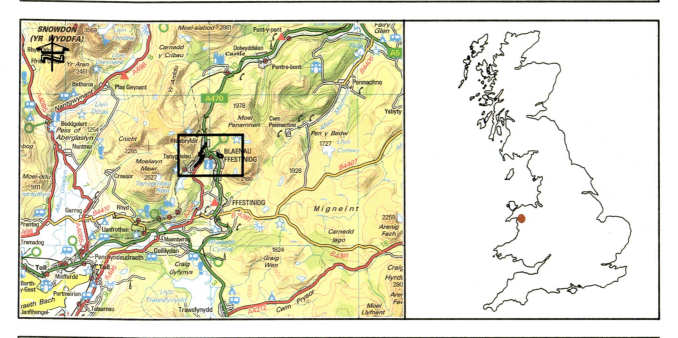

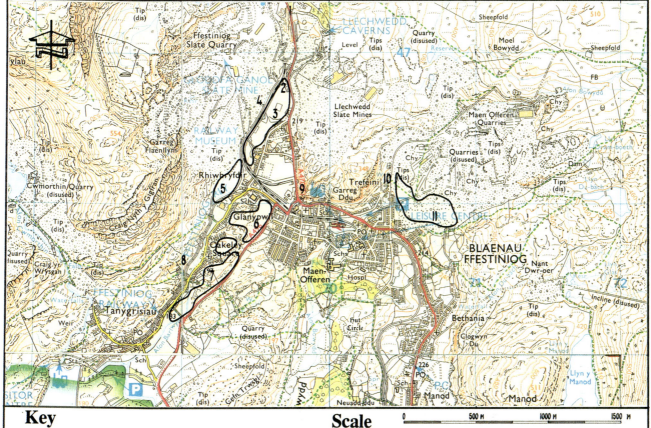

Key

Scale

| 0 | 500 M | 1000 M | 1500 M |

1. **A470(T) north**
2. **Former slate loading sidings**
3. **Glan Y Don tip removed from this site in 1975 Factory built in 1994**
4. **Possible location for slate loading sidings**
5. **Area filled for development under Glan Y Don scheme**
6. **Area filled for development under Glan Y Don scheme**

7. **Area filled for development under Glan Y Don scheme**
8. **Narrow gauge steam railway**
9. **Natural rock bluff**
10. **Natural rock bluff**
11. **Fotty tip partially regraded in 1986**

1. Blaenau Ffestiniog developed in the centre of the slate workings.

2. Slate waste tips dominate the skyline of the town centre.

3. Waste tipping alongside the A470 is extending the area affected by quarrying.

4. The Glan y Don slate waste tip filled the valley floor between the trunk road and railway. The scale of historic waste tipping is clearly visible.

5. Reclamation opened the access to the town and provided space for recreation and development. The town still has a rail link which could transport secondary aggregates from the active quarries.

Nantlle Valley, Gwynedd

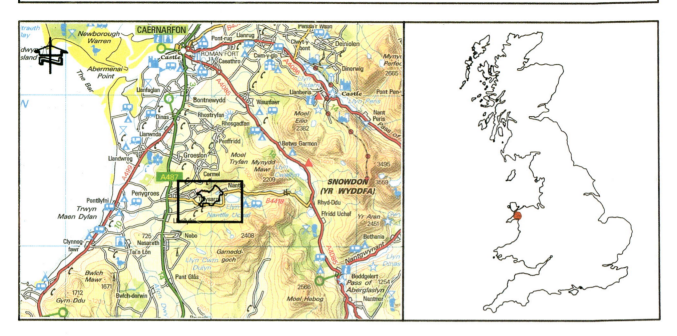

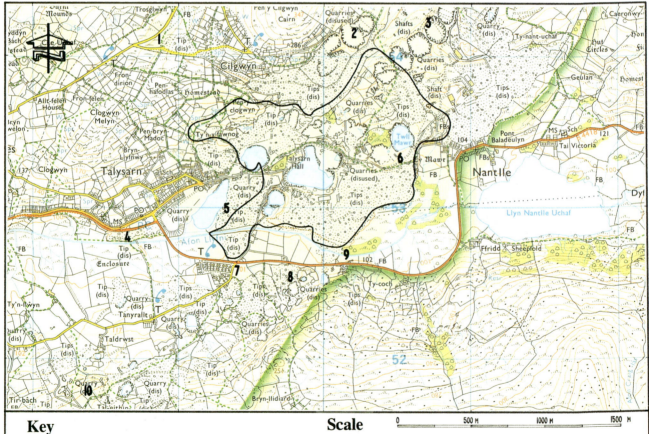

Key

Scale

1. **Trosglwyn tip**
2. **Cilgwyn waste disposal site**
3. **Pen Yr Orsedd quarries (active)**
4. **Coed Madog reclamation scheme**
5. **Talysarn reclamation scheme**
6. **Twll Ballast**
7. **Road access from Fron Heulog to B4418**

8. **Ty Mawr East quarry**
9. **Access for proposed waste removal**
10. **Fron Heulog quarry (active)**
Boundary of Snowdonia National Park shown thus:
Dorothea Quarries complex shown thus: ———

1. *The conservation of industrial archaeological interest and wildlife habitats were among the reasons for the refusal of planning permission for waste extraction at Dorothea Quarry.*

2. *Industrial wastes were tipped into Twll Ballast between 1963 and 1974. The landslip is a reminder of the need for periodic inspection of quarry faces.*

3. *The Talysarn scheme involved extensive reshaping of the slate waste tips around the quarry hole, to form a shallow lake. Further vegetation development work would divide the expanses of grass and enhance the site.*

Allt Ddu, Llanberis, Gwynedd

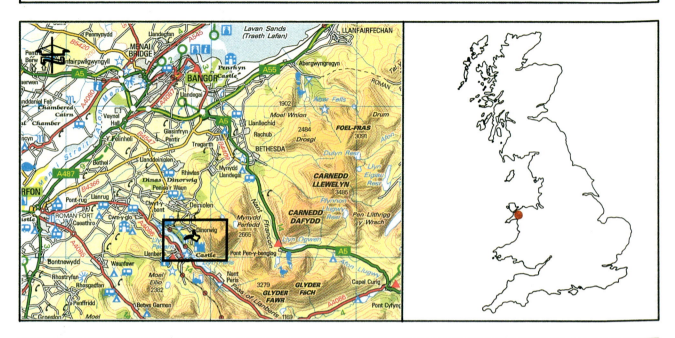

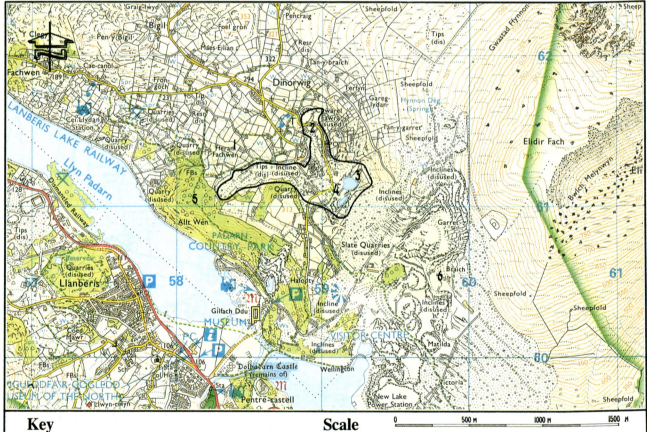

Key

Scale

1. **Small enclosure pattern**
2. **Public road on edge of quarry**
3. **Quarry face used for climbing**
4. **Public road on edge of quarry**
5. **Oak woodland SSSI**
6. **Dinorwig quarry complex**

Boundary of Snowdonia National Park shown thus:

1. The slate waste surface left by partial removal of the tip. After surface crushing by grid roller, grass was sown directly on to the slate waste.

2. Trees were planted into a layer of crushed, screened slate waste. The polythene mulch helps to conserve water in the root zone.

3. The scheme has improved the environment of these houses. In the background the massive tips of the Dinorwic complex remain.

Ballachulish, Highland

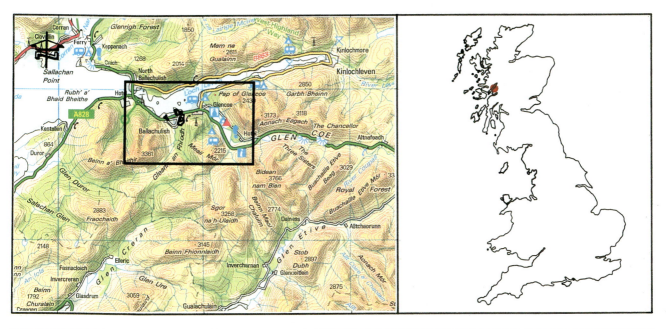

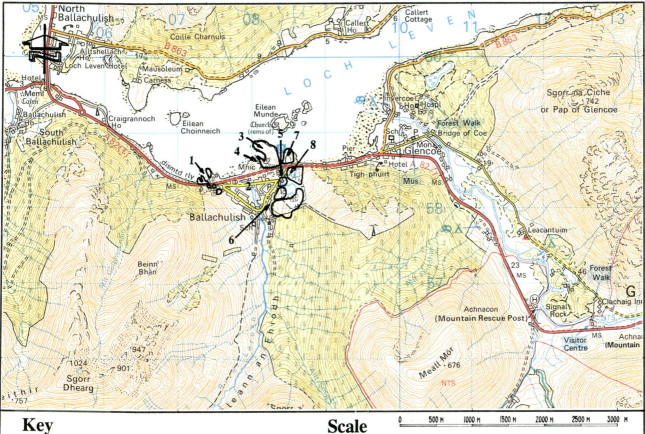

Key

Scale

1. **Old quarry used as highways depot**
2. **A82(T) realigned**
3. **Tips reduced and grassed**
4. **Hotel and leisure development**
5. **New tourist information centre**
6. **Extensive tree planting on regraded slate waste**
7. **Tip reduced, inlet depth reduced**

8. **Old incline retained**

1. The reclamation of slate waste tips at Ballachulish was the key element in a strategy to revitalise the village.

2. High rainfall contributes to the continued growth of grass and trees.

Hodge Close, Tilberthwaite

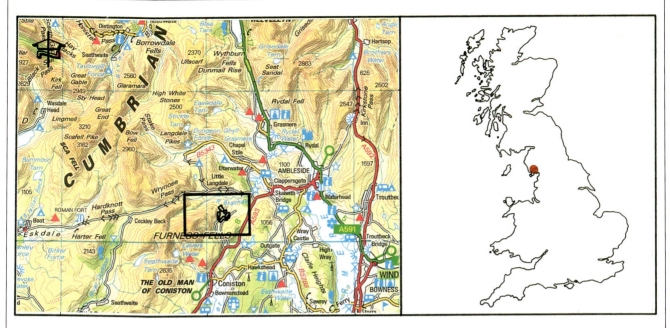

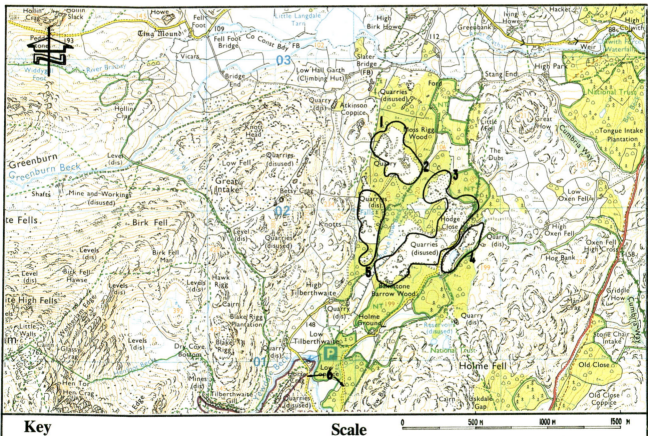

Key

Scale

1. **Moss rigg quarry (dormant)**
2. **Prominent modern building**
3. **Extensive natural woodland screens workings**
4. **Part flooded quarries used for climbing, abseiling and sub-aqua**
5. **Small scale working of slate continues**
6. **Single track roads restrict access from A593**

1. *The road leading to this site is narrow and winding, and so the extraction of slate and slate waste on a large scale would be impractical and unwelcome. Extensive woodland and colonisation of the waste softens the outline of the tips.*

PHOTOGRAPH COURTESY OF LAKE DISTRICT NATIONAL PARK AUTHORITY

2. *The site is used for recreational purposes including rock-climbing. Visitor pressure could limit the process of recolonisation.*

Welsh slate museum, Llanberis, Gwynedd

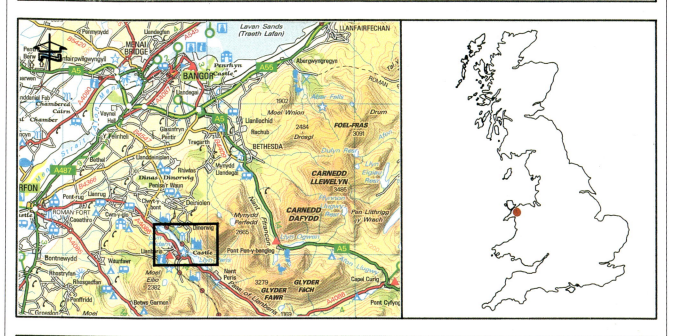

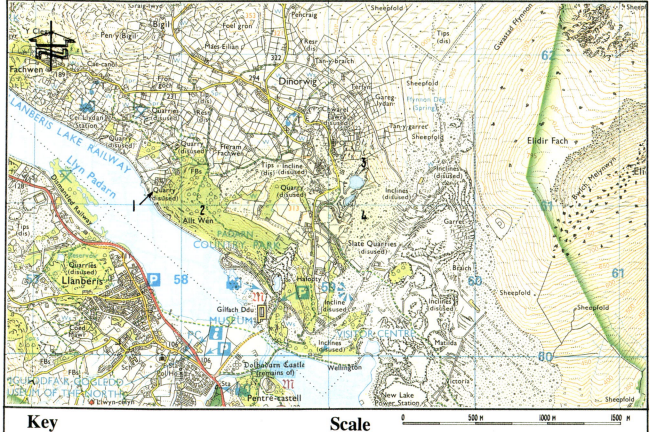

Key

Scale

1. **Viewing platform**
2. **Vivian quarry**
3. **Dinorwig quarry complex**
4. **Pumped storage power station**
    Boundary of Snowdonia National Park shown thus:

1. The Welsh Slate Museum provides the focus for a range of recreational activities.

2. Quarrying history and countryside trails are combined at this site.

Appendices and References

Appendix 1
Organisations consulted

Appendix 2
Slate working sites visitied during the review

Appendix 3
Land reclamation schemes

Appendix 4
Organisations providing financial assistance

Appendix 1

Organisations consulted *(excluding bodies which did not reply)*

Quarry operators

Cumbria:
Burlington Slate Ltd, Kirkby in Furness
Kirkstone Green Slate Ltd, Skelwith Bridge

Wales:
Alfred McAlpine Slate Products Ltd, Bethesda
Ffestiniog Slate Quarries Ltd, Blaenau Ffestiniog
Greaves Welsh Slate Co, Blaenau Ffestiniog
Redland Aggregates Ltd, Blaenau Ffestiniog
Wincilate Ltd, Aberllefenni

Cornwall:
RTZ Mining and Exploration Ltd, Delabole
Realstone Ltd, Bolehill Quarry

Overseas:
Anthony Dally and Sons Inc, Pennsylvania

National Organisations:
British Aggregate Construction Materials Industries,
    London
The Institute of Quarrying, Nottingham

Local government

County Councils:
Clwyd County Council, Director of Development
    and Tourism
Cornwall County Council, County Planning
    Department
Cumbria County Council, County Planning Officer
Devon County Council, County Engineering and
    Planning Officer
Dyfed County Council, County Planning Department
Gwynedd County Council, Deputy County Archivist
    and Museums Officer
Gwynedd County Council, National Park Officer
Gwynedd County Council, County Planning Officer
Powys County Council, County Planning Officer
Powys County Council, County Archives Officer

Regional Councils:
Central Regional Council
Grampian Regional Council
Highland Regional Council
Strathclyde Regional Council
Tayside Regional Council

District Councils:
Aberconwy Borough Council
Arfon Borough Council
Argyll and Bute District Council
Banff and Buchan District Council
Caradon District Council
Dumbarton District Council
Eden District Council
Glyndwr District Council
Gordon District Council
Kincardine and Deeside District Council
Lochaber District Council
Meirionnydd District Council, Chief Executive's
    Department
Meirionnydd District Council, Environmental Health
    Department
Meirionnydd District Council, Technical Services
    Department
Moray District Council
North Cornwall District Council
Restormel District Council
Stirling District Council, Policy and Strategic Planning
South Lakeland District Council

Government, statutory and non-statutory organisations

Welsh Office:
Welsh Office Landscape Advisor
Planning Division

Regional Aggregates Working Parties:
North Wales Working Party on Aggregates
South West Regional Aggregates Working Party
Northern Region Working Party on Aggregates

National Park Authorities:
Gwynedd County Council, Snowdonia National
    Park Officer
Lake District National Park

HM Inspectorate of Quarries, Land Reclamation
    Division London, North West Region,
    South West Region

Welsh Development Agency

Scottish Enterprise, Edinburgh

Scottish Office Environment Department

National Rivers Authority

River Purification Boards:
Clyde River Purification Board
Forth River Purification Board
Highland Rivers Purification Board
Tay River Purification Board

Tourist Boards:
Cumbria Tourist Board
Scottish Tourist Board
Wales Tourist Board

English Nature, Peterborough

Countryside Commission, Cheltenham

Scottish Natural Heritage:
Scottish Natural Heritage, Fort William
Scottish Natural Heritage, Golspie, Sutherland

Countryside Council for Wales

Royal Society for the Protection of Birds: Sandy, Bedfordshire
RSPB, Dolgellau
RSPB, Newtown

National Museum of Wales:
Welsh Slate Museum, Llanberis

Industrial Archaeology:
Association for Industrial Archaeology, Aberystwyth
Cadw: Welsh Historic Monuments
National Association of Mining History Organisations
Northern Mine Research Society
RCAHMS, Edinburgh
RCAHMW, Aberystwyth

Countryside Protection Groups:
Friends of the Lake District, Kendal
Campaign for the Protection of Rural Wales, Welshpool
Scottish Wildlife Trust, Stirling
Cumbria Wildlife Trust, Ambleside
Devon Wildlife Trust, Exeter

Appendix 2

Slate working sites visited during the review

England

Cornwall

Caradon District
St Neot Quarry, Carnglaze.   Tourist Centre

North Cornwall District

Delabole	Active
Merryfield/Lower Tynes	Active
Higher Tynes	Active
Trebarwith Road	Dormant
Prince of Wales	Dormant
Bolehill	Active
Trevillet	Active
Trecarne	Active (intermittent)
Jeffries Pit	Reclaimed
Bowithick	Refuse disposal site (completed)
Bowithick East	Filled and regraded

Cumbria

Lake District National Park

Petts	Active
Hodge Close	Active
Elterwater	Active
Low Brandy Crag	Active
Thrang Crag	Dormant
Honister Pass	Dormant
Quay foot, Borrowdale	Dormant

South Lakeland District
Kirkby   Active

Scotland

Central Region, Stirling District
Aberfoyle   Dormant (waste extraction)

Dumbarton District
Luss   Dormant

Highland Region, Lochaber District
Ballachulish   Reclaimed

Strathclyde Region, Argyll and Bute District
Ellanabeich   Dormant

Wales

Clwyd
Glyndwr District

Moel y Faen, Horseshoe Pass	Dormant, Waste extraction
Berwyn, Horseshoe Pass	Dormant

Dyfed

Preseli Pembrokeshire District

Glogue	Dormant
Rosebush	Dormant
Gilfach	Dormant
Temple Druid	Dormant

Gwynedd

Aberconwy Borough

Cwm Machno	Reclaimed
Chwarel Ddu, Dolwyddelan	Reclaimed
Tyn y Bryn, Dolwyddelan	Dormant

Arfon Borough

Penrhyn Quarry, Bethesda	Active
Allt Ddu and Chwarel Fawr, Deiniolen	Reclaimed
Vivian, Llanberis	Conserved
Dinorwig, Llanberis	Part conserved
Marchlyn, Llanberis	Dormant
Glanyrafon, Rhyd Ddu	Dormant
Coed Madoc, Talysarn	Reclaimed
Talysarn, Talysarn	Reclaimed
Dorothea, Nantlle	Dormant
Ty Mawr East, Nantlle	Waste extraction
Twll Ballast, Nantlle	Dormant
Taldrwst and surrounding quarries, south of Nantlle	Dormant
Fron Heulog, Nantlle	Waste extraction

Meirionnydd District

Cwm Orthin, Blaenau Ffestiniog	Active
Rhosydd, Blaenau Ffestiniog	Dormant
Wrysgan, Blaenau Ffestiniog	Dormant
Gloddfa Ganol Complex, Blaenau Ffestiniog	Active
Llechwedd Complex, Blaenau Ffestiniog	Active
Fotty tip, Blaenau Ffestiniog	Reclaimed
Glan y Don tip, Blaenau Ffestiniog	Reclaimed
Abercwmeiddau, Upper Corris	Reclaimed
Aberllefenni, Corris	Active
Braichgoch, Corris	Reclaimed

Appendix 3

Land reclamation schemes

1. Jeffries Pit and Prince of Wales Quarry
   Nr Tintagel, Cornwall.
   NGR SX 073863
   Implemented 1969
   and 1989-90

Objectives	To provide car parking and picnic facilities, and an attractive walk at Jeffries Pit. To provide interpretive trails around the abandoned Prince of Wales quarry.
Works	Clearance of fly tipped refuse and part filling of quarry. Surfacing of parking area in unbound aggregate. Waymarking and interpretive signing. Footpaths and steps, safety fencing.
Area	2.5 hectares at Jeffries Pit
Works cost	Not available.
Current Status	The sites are well used. The Prince of Wales Quarry Engine House was restored by enthusiasts in 1976. The Heritage Coast Service manage the sites and published a leaflet interpreting the quarry, engine house and the wildlife which has colonised much of the site.

2. Vivian Quarry, Llanberis, Gwynedd
   See Case Study 7.

3. Coed Madog, Talysarn, Gwynedd
   NGR SH 487589
   1972

Objectives	The first step in a programme to improve a village surrounded by slate waste tips and quarries abandoned as recently as 1970. To open up the views and access to Snowdonia from the west.
Works	Clearance of the derelict railway yard. Reduction and grassing of low level slate waste tip. Creation of parking and football areas adjoining village, and access to river side. Conversion of station building to meeting room.
Area	8.7 hectares

Works cost	£50,000.
Current Status	Parking area well used. Football area neglected but new play area developed. Low-level tip subsequently incorporated into Talysarn scheme (see 7).

4. Pantdreiniog Quarry, Bethesda, Gwynedd
   NGR SH 624671
   1972

Objectives	Removal of danger from a disused quarry close to the town and school. Provision of recreational and development land. Removal of slate waste tips extending into a residential area of the town.
Works	Regrading the waste tip and filling of the quarry. Establishment of a grassed play area.
Area	10.2 hectares
Works cost	Not available
Current Status	The recreational area is considered surplus to local needs. The Borough Council may promote housing development on part of the site (Hughes D. pc).

5. Braichgoch, Corris, Gwynedd
   NGR SH 749079
   Designed 1972-75
   Implemented 1975-78

Objectives	Elimination of danger from unsafe tip retaining walls. Replacement of severely sub-standard trunk road and provision of land for rural development. Visual improvement by treatment of slate waste tips on valley floor and valley side.
Works	Regrading of 650,000 m³ of slate waste to remove retaining walls and tips, form a development platform and provide the formation for a new 7.3m wide trunk road. Crushing slate waste for use as sub-base, drainage backfill and surface dressing before grass seeding. Construction of the 1 km long new trunk road, and access to the development area.

Service diversions, provision of new foul sewer and surface water drainage system. Provision of a new footpath separated from the road.
Planting blocks of native trees.

Area	15.3 hectares
Works cost	£700,000
Current Status	A popular and successful Craft Centre was built shortly after the reclamation scheme. The trunk road has improved communications. The trees have generally grown well or extremely well and tree seedlings are rapidly invading the slopes as the grass declines. The trees received one weed control and fertiliser treatment in 1983, but no other maintenance has been carried out.

6. Glan y Don, Blaenau Ffestiniog, Gwynedd
    NGR SH 696466
    Designed 1972-75
    Implemented 1975-77

Objectives	Removal of slate waste tip at rear of houses and creation of open space between A470(T) and British Rail entrance to the town.
Provision of fill for reuse in preparing development land.
Works	Excavation of tip, compaction of waste and filling of land for development. 750,000m³ of slate waste was moved.
River diversion.
Diversion underground of many unsightly overhead services.
Widening a sub-standard public road.
Contouring a fill area for the possible use of the Ffestiniog railway.
Construction of playing field and new road and river bridges.
Dressing of surface with fine mill waste, before grass seeding.
Planting of native trees in beds of subsoil.
Area	11.9 hectares
Works cost	£600,000
Current Status	Lack of initial aftercare led to poor grass growth and regression, but tree growth has been good. Public open space little used, and under consideration for development. Development land taken up for housing. Playing fields well used.

7. Talysarn Tip, Talysarn, Gwynedd
    NGR SH 493532
    1978-79

Objectives	Removal of prominent tip which dominated the village and prevented vehicular access to some houses.
Reduction of danger presented by sheer quarry face and flooded pit at the rear of some houses.
Removal of unsafe, crumbling tramway arch over main access.
Provision of small development plots.
Works	Demolition of tramway arch.
Regrading of tip and filling of quarry.
Partial filling of further flooded quarry to create shallow lake.
Reduction in height of Coed Madog tip to provide further fill (see 3).
Movement of 900,000 m³ of slate waste in total.
Construction of playing field, road access and vehicle turning circle.
Surface crushing and grass seeding.
Area	17.0 hectares
Works cost	£650,000
Current Status	Lack of aftercare has led to poor grass growth and regression. Gorse has invaded the open grass sward. 'Sheltered' bungalows were built on one development plot. The Borough Council may incorporate the site into a proposed country park (Hughes D. pc).

8. Llanberis Industrial Estate/Y Glyn, Llanberis, Gwynedd
    NGR SH 573608
    1982

Objectives	Provision of industrial development land.
Provision of safe public access to lakeside.
Removal of hazard due to submerged slate waste.
Works	Regrading of slate waste tips, and compaction for development.
Provision of road access and infrastructure to development area.
Minimal clearance, levelling and reshaping within low level lakeside tips, conserving mature tree cover.
Formation of footpaths, sheltered lagoons and 'beaches'.
Dressing 'beaches' and lagoons with rounded shingle, for public safety.

Area	13.5 hectares at industrial estate; 2.5 hectares at Y Glyn
Works cost	£45,000 (Y Glyn recreation area only).
Current Status	The development area has been taken up for industry. The recreation area was awarded an RICS/Times Conservation Award for its sensitive approach and low cost. The area is well-used by the public.

9. Allt Ddu and Chwarel Fawr, Llanberis, Gwynedd.
See Case Study 4

10. Ballachulish Quarries, Nr Glencoe, Highland.
See Case Study 5

11. Fotty Tip, Blaenau Ffestiniog, Gwynedd
NGR SH 705462
Designed 1980
Implemented 1982-87

Objectives	Prevention of severe flooding of the town by creating space to construct a flood control system. Removal of waste tip adjacent to houses. Visual improvement from town centre.
Works	Improvements to the river culvert and channel through the town centre. Construction of a new floodwater collecting system at the base of the tips. Movement of 200,000m³ of slate waste to remove a relatively small tip near houses, and regrade part of a large tip to create space for the flood prevention works. Construction of slate lined channels to redefine the main incline which once served the quarries. Tree planting in slate waste and underlying material. Note: The original proposal to regrade a much larger area of tips within the confines of the active quarry, was not implemented because the proposals were considered by some people to involve too great a change in the landscape.
Area	4.4 hectares
Works cost	Not available.
Current Status	The planted areas are developing slowly. Growth of trees planted in the underlying soil is more rapid. There has been no recurrence of flooding.

12. Glyn Rhonwy Quarries, Llanberis, Gwynedd
NGR SH 5660 (Grid square)
Implemented 1986-93

Objectives	Clearance of temporary buildings. Provision of industrial development land. Retention of existing tree cover.
Works	Dismantling and demolition of buildings. Preparation of development platforms on parts of the quarry-tip complex. Use of slate waste for minor regrading, and as fill off-site. Construction of access roads and services. Landscape works associated with development. Note: Various proposals for leisure-related uses of the quarry and tips complex have been put forward but none has yet been developed (Hughes D. pc). 16 ha of development plots now prepared.
Area	114 hectares (site total)
Works cost	Not available - under construction.
Current Status	One large high-technology factory is now in operation. Further construction contracts were in progress in 1994.

13. Chwarel Ddu, Dolwyddelan, Gwynedd
NGR SH 723522
Designed 1983-84
Implemented 1988-90

Objectives	To create a small car park and picnic site in conjunction with improvements to the A470 (T), serving visitors to Dolwyddelan castle. To remove unsightly slate waste tips.
Works	Use of slate waste as fill for road construction, and regrading within the site. Peat and soil importation from the road works. Grass seeding and planting. Picnic facilities. Safety walling and fencing around the flooded quarry, alongside the scheme.
Area	1.9 hectares
Works cost	Not available - scheme integrated with A470 works.
Current Status	In use and managed by the Snowdonia National Park Authority. Grass growth is good but overgrazed in parts. Tree growth has been slow during the first few years.

14. Cwm Penmachno, nr Betws y Coed, Gwynedd
    NGR SH 752472
    Designed 1982
    Implemented 1983

Objectives	Removal of collapsing tramway arch over main access track and popular footpath.
	Removal of dangerous structures.
	Clearance of fly tipping.
	Continued public access to main workings and hillside.
Works	Demolition of arch and structures (after recording by County Archivist).
	Regrading of adjacent tips (23 200 cu.m of waste).
	Removal of fly tipped cars, and partial filling of lowest quarry.
	Spreading of local subsoil and road verge clearings.
	Grass seeding (2 ha).
Area	4.9 hectares
Works cost	£20,000
Current Status	This minor scheme allows safer access to the 29 ha quarry complex, for school and outward bound groups. The grass area is owned and managed for grazing by a local farmer, and has retained a dense sward.

15. Abercwmeiddaw, Upper Corris, Gwynedd
    NGR SH 745094
    Designed 1983-85
    Implemented 1989-90

Objectives	Removal of danger from unstable tip retaining wall near houses and alongside a public road.
	Provision of land for small scale development.
	Provision of improved access to quarry for waste disposal.
Works	Regrading of the hillside tip and

removal of the retaining wall.
Construction of single track access for heavy vehicles.
Construction of access ramp to quarry floor.
Regrading of riverside tip over new river culvert to form new road access to the A487(T).
Selection and placing of fine slate on benches, and tree planting.

Area	8.0 hectares
Works cost	£550,000
Current status	Waste disposal proposals now abandoned in light of modern environmental protection standards. Good tree survival and initial growth. Natural colonisation of slate surface beginning, but the tip remains prominent due to colour and smooth texture. Development plots not yet taken up.

16. Trosglwyn Tip, Carmel, Nr Caernarfon, Gwynedd
    NGR SH 541490
    Designed 1992

Objectives	Removal of an isolated small tip close to houses.
	Restoration of vegetation to match the adjoining common land.
Proposed Works	Excavation of the tip and removal for use as cover in a nearby landfill site.
	Cultivation of the underlying soil, fencing to exclude stock, and allowing wind-blown grass, heather and gorse to establish.
Area	3.0 hectares
Current status	Following consultations the scheme was granted planning permission, and is ready to supply cover to the landfill when required.

Appendix 4

Organisations providing financial assistance

As the arrangements and scope of grant schemes are liable to revision, readers are advised to obtain current information from the organisations listed.

A. Land reclamation

England
Arrangements for the operation of grant-aid for derelict land reclamation were not available at the time of publication. The grant will be administered by the urban regeneration agency English Partnerships.

Scotland
Grant-aid for derelict land reclamation is administered by local enterprise companies. Details of the current grant arrangements may be obtained from:

Scottish Enterprise
120 Bothwell Street
Glasgow G2 7JP
Tel: 0141-248-2700
Fax: 0141-221-1983

Wales
Grant-aid for derelict land reclamation and environmental improvement is administered by the Welsh Development Agency. Details of the current grant arrangements may be obtained from:

Director of Land Reclamation
Welsh Development Agency
Pearl House
Greyfriars Road
Cardiff CF1 3XX
Tel: 01222 222666
Fax: 01222 390752

B. Freight facilities

The Rail Freight Facilities Grant Scheme and Inland Waterways Freight Facilities Grant Scheme are administered by:

Railways Directorate
Department of Transport
2 Marsham Street
London SW1P 3EB
Tel: 0171-276-4834 or 3154

Scottish Office Industry Department
New St Andrews House
St James Centre
Edinburgh EH1 3SZ
Tel: 0131-244-4140

Welsh Office
Government Buildings
Ty Glas Road
Llanishen
Cardiff CF4 5PL
Tel: 01222-761456 ext 5282

C. Works for nature conservation and landscape enhancement

The principal schemes for these objectives are administered by:

English Nature
Northminster House
Peterborough PE1 1UA
Tel: 01733-340345

Countryside Commission
John Dower House
Crescent Place
Cheltenham
Gloucestershire GL50 3RA
Tel: 01242 521381
Fax: 01242 226027

Countryside Commission for Scotland
Battleby
Redgorton
Perth PH1 3EW
Tel: 01738 27921

Countryside Council for Wales
Plas Penrhos
Ffordd Penrhos
Bangor
Gwynedd LL57 2LQ
Tel: 01248 355141

Details of the relevant schemes, and the address of the relevant local or regional office, can be obtained from the headquarters addresses given. Local authority planning departments or countryside departments administer some grants eg the Woodland Grant Scheme, on behalf of the national bodies, and are also able to advise on the scope and availability of other grants.

D.   New technology and business

The government supports and encourages the development and adoption of new technology and new environmental management systems.

The Environmental Technology Innovation Scheme, ETIS, is administered by:

ETIS Office
Department of the Environment
Room B351
Romney House
43 Marsham Street
London SW1P 3PY
Tel: 0171-276-8318
Fax: 0171-276-8333

The DTI's Environmental Management Options Scheme, DEMOS, is administered by:

DTI Environment Unit
3rd Floor
151 Buckingham Palace Road
London SW1W 9SS
Tel: 0171-215-1065

Information concerning current assistance, both financial and technical, for business development can be obtained from the local 'enterprise centre' or local authority business support unit.

References

A.   Literature

Al-Gosaibi (1985) Bioengineering properties of soil stabilisers and mulches. Ph.D. Thesis, University of Liverpool.

Anon (1986) Cumbria and Lake District Joint Minerals Local Plan. Written Statement. Deposit Draft. March 1986.

Anon (1987) Lake District National Park Visitor Survey. Lake District National Park Authority (unpublished).

Anon (1988) Council Directive 89/106/EEC of 21 December 1988 on the approximation of laws, regulations and administrative provisions of the Member States relating to construction products. Official Journal of the European Communities No L 40, 11.2.1989.

Anon (1990a) British Standard 1377: Methods of test for soils for civil engineering purposes. Part 9 'In-situ tests'. British Standards Institution.

Anon (1990b) This Common Inheritance. CM 1200:1992. HMSO.

Anon (1993) Major secondary aggregate proposal refused permission in Gwynedd, Mineral Planning, 54, p20-22. March, 1993.

Anon (1993b) Council Directive 93/37/EEC of 14 June 1993 concerning the co-ordination of procedures for the award of public works contracts. Official Journal of the European Communities No. L 199/54. 9.8.93.

Armstrong, R A (1981) Field experiments on the dispersal, establishment and colonization of lichens on a slate rock surface, Environmental and Experimental Botany, 21, No. 1, 115.

Arup Economics and Planning (1991) Occurrence and utilisation of mineral and construction wastes Final Report. London: Arup Economics and Planning.

Association of Industrial Archaeology (1991) Industrial Archaeology: Working for the Future.

WS Atkins Environment (1992) Twll Ballast Rehabilitation: Executive Summary (Report to Arfon Borough Council).

Blunden, J R (1975) The mineral resources of Britain: A study in exploitation and planning. Longman.

Blunt, S (1991) Practical techniques for the revegetation of derelict land. M.Sc. Thesis, University of Liverpool.

Bradshaw, A D and Chadwick, M J (1980) The Restoration of Land. Blackwell.

Bragg, M (1983) Land of the Lakes. Hodder.

Broad, K F (1979) Tree Planting on Man-made Sites in Wales. Forestry Commission.

Building Research Establishment (1993) Efficient Use of Aggregates and Bulk Construction Materials (Report to DOE).

Building Research Establishment (1994) Use of Wastes and recycled materials as aggregates: standards and specifications (in preparation).

Cameron A D (1989) Honister Slate Mine. The Mine Explorer (Journal of Cumbria Amenity Trust), 3.

Clarke, G. and Harding-Thompson, W. (1938) Lakeland landscape.

Countryside Commission (1984) A Better Future for the Uplands. CCP 162.

Countryside Commission (1993) Landscape Assessment - New Guidance. CCP 423.

Crompton, C T (1967) The treatment of waste slate heaps, Town Planning Review, 37, 291-304.

Cuthbertson, D (1983) Climber's Guide to Greag Dubh and Craig Barns. Scottish Mountaineering Club.

Daily Post (1993) Newspaper article 22.3.93.

Department of the Environment (1988a) Minerals Planning Guidance. General Considerations and the Development Plan System. MPG 1. London: HMSO.

Department of the Environment (1988b) Minerals Planning Guidance. Applications, Permissions and Conditions. MPG 2. London: HMSO.

Department of the Environment (1989). Minerals Planning Guidance. Guidelines for Aggregates Provision in England and Wales. MPG 6. London: HMSO.

Department of the Environment (1990) Welsh Office (1991) Planning Policy Guidance, Archaeology and Planning. PPG 16. London: HMSO.

Department of the Environment (1991a) National collation of the results of the 1989 Aggregate Minerals Survey. Department of the Environment.

Department of the Environment (1991b) Derelict Land Grant Policy. DLGA 1. Department of the Environment.

Department of the Environment (1992a) Monthly statistics for building material and components. Dec 1992 HMSO.

Department of the Environment (1992b) The Operation of the Derelict Land Grant Scheme. Advice Note 3. unpublished.

Department of the Environment (1992c) The Urban Regeneration Agency. A Consultation Paper Issued by the Department of the Environment, July 1992.

Department of the Environment (1993) Guidelines for Aggregates Provision in England and Wales. Revision of MPG 6. Draft Consultation Document. London:HMSO.

Department of the Environment (1994a) The Reform of Old Mineral Planning Permissions. Consultation paper. DOE.

Department of the Environment (1994b). Minerals Planning Guidance. Guidelines for Aggregates Provision in England. MPG 6 London: HMSO.

Department of the Environment/Secretary of State for Scotland/Secretary of State for Wales (1976) Aggregates: The Way Ahead (The Verney Report). London: HMSO

Department of the Environment/Welsh Office (1985) Circular 1/85. Use of Conditions. DOE.

Department of the Environment/Welsh Office (1991) Circular 14/91. Planning and Compensation Act. DOE.

Department of the Environment/Welsh Office (1993) The Control of Noise at Surface Mineral Workings. MPG11 DOE.

Department of Transport (1991) Freight Facilities Grant. HMSO.

Department of Transport (1992) The Government's response to the SACTRA report on assessing the environmental impact of road schemes. Department of Transport.

Derbyshire County Council (1992) Derbyshire Stone Slate Roofs. Guidance for Owners of Historic Buildings. Derbyshire County Council.

English Estates, Scottish Development Agency, Welsh Development Agency, Industrial Development Board of Northern Ireland, Development Board for Rural Wales, Highlands and Islands Development Board, (1986) Planning and site development: industrial and commercial estates. Thomas Telford 1986.

English Nature (1991) Conserving our heritage of rocks, fossils and landforms. English Nature.

English Nature (1992) Earth Science Conservation for the Mineral Extraction Industry. English Nature.

Finegan B G, Harvey H G, Humphreys R N (1983) Vegetation succession in abandoned chalk quarries. In Reclamation '83 Conference Papers, 472-482, Industrial Seminars Ltd.

Geoffery Walton Practice (1991) Handbook on the Design of Tips and Related Structures. London: HMSO.

Gilchrist, A R (1981) Ballachulish, Its Early Grandeur Restored. Landscape Architecture 748-751 November 1981.

Gutt W, Nixon P J, Smith M A, Harrison W H, Russell A D (1974) A survey of the locations, disposal and prospective uses of the major industrial by-products and waste materials. In Building Research Establishment Current Paper CP 19/74.

Harries-Rees K (1991) Slate market split, Industrial Materials, May 1991, 44-55.

Harte J D C (1985) Landscape, Land Use and the Law. Spon.

Johnson G, Johnson C F, editors (1992) Snowdonia and Its Coast. Series publication. South Gwynedd Leader Network.

Land Use Consultants (1990) Environmental Improvement Study: Horseshoe Pass. Summary Report (to Glyndwr District Council, Clwyd County Council, WDA and Welsh Tourist Board).

Land Use Consultants (1992) Amenity Reclamation of Mineral Workings (Report to the DOE) HMSO.

Lewis M J T and Denton J (1974) Rhosydd Slate Quarry. Cottage Press.

D Lovejoy and Partners (1988) The Dorothea - a quarry park (report to Gwynedd County Council and others).

Luke A and Macpherson T (1983) Direct seeding: a potential aid to land reclamation in Central Scotland. Arboricultural Journal 7, 287-299.

McFadzean A (1986) The Mine Explorer Journal of the Cumbria Amenity Trust.

McGowan I J (1982) Land renewal at Ballachulish. Landscape Design 138, 11-12.

Ministry of Works (1947) The Welsh Slate Industry. HMSO.

Moffat A and McNeill J (1994) Reclaiming disturbed land for forestry. Bulletin 110. HMSO London.

Moon R (1992) What the forecast means for a major exporting county: a county view. Minerals Planning 49, December 1992, 12-16.

National Trust (1993) Members' Newsletter Spring 1993.

Nature Conservancy Council (1989) Guidelines for the Selection of Biological SSSIs. Nature Conservancy Council.

Nature Conservancy Council (1990) Handbook for Phase 1 Habitat Survey. NCC 1990.

Parkman Consulting Engineers, Chester (1991) A5 Bethesda Bypass. Geotechnical desk study report Vol 1. Final draft. (Summary)

Peacock (1983) Slate waste - properties and prospective uses. M.Sc. Thesis, University of Durham.

Ratcliffe D A (1974) Ecological effects of mineral exploitation in the United Kingdom and their significance to nature conservation. Proceedings of the Royal Society of London, A Series 339, 355-372.

Richards A J (1991) A Gazetteer of the Welsh Slate Industry. Gwasg Carreg Gwalch.

Richards, Moorehead and Laing Ltd (1992) Ty Mawr East planning application: supporting statement (unpublished).

Robinson, Jones Partnership (1987) Working with Nature Welsh Development Agency Cardiff.

Roy Waller Associates Ltd (1991) Environmental effects of surface mineral workings (Report to the DOE) HMSO London.

Royal Society for Nature Conservation/The Wildlife Trusts Partnership (1992) Blasts from the Past. Royal Society for Nature Conservation.

Scottish Office (1992) Scottish Vacant and Derelict Land Survey 1990.

Scottish Office (1994) Land for Minerals Working NNPG 4. Scottish Office 1994.

Sheldon J C (1975) The reclamation of slate waste. Nature in Wales 14, RP 160-168.

Stephens Associates (1988) Slate Quarrying in the Lake District - a cause for concern. Friends of the Lake District.

Stevens T (1987) Llechwedd Slate Caverns. Leisure Management 7, No.4, 27-29.

Vernon R (1989) Conservation of mining sites in the Gwydir Forest area of Snowdonia National Park. Industrial Archaeology Review 12, No.1, 79.

Welsh Development Agency (1990) Survey of Derelict Land in Wales 1990. (Unpublished)

Welsh Industrial Archaeology Panel (1992) Guidelines for procedures and standards for archaeological investigation and field work on industrial archaeological sites in Wales. Second Draft.

Welsh Office (1990) Roads in Upland Areas: A Design Guide. Welsh Office Highways Directorate.

Welsh Office (1994) Guidelines for Aggregates Provision in Wales. Welsh Office (in preparation).

Williams G (1992) Please don't try to move our slate tip. Daily Post, 14 December 1992.

Williams M (1991) The Slate Industry. Shire Publications Ltd.

Williamson I A (1988) Report on the Position of the Tip Complex at Petts Quarry, Kirkstone, nr Ambleside. (Unpublished)

Wright R (1992) Landfill and our Geological Heritage. Waste Planning, No.4, 10-13.

B. Personal Communications

Information provided by the following individuals or organisations is referred to in the text. Where the individual represented an organisation this is shown.

Barker J.	Planning department, Gordon District Council
Brownlee M.	Burlington Slate Ltd, Kirkby Quarry, Cumbria
Byrne C.	Resident Engineer, Fotty and Bowydd Reclamation Scheme
County Planning Officer	Dyfed County Council
County Planning Officer	Powys County Council
Davies K.	Senior Planning Officer, Countryside Council for Wales
Eccleston R.	British Sub Aqua Federation
English Nature	Regional Office, Windermere
Fecitt N.	Managing Director, Kirkstone Green Slate Quarries Ltd, Cumbria
Fidler J.	Head of Architectural Conservation, English Heritage
Grenter S.	Industrial Archaeologist, Clwyd County Council
Griffiths D G.	Director of Land Reclamation, Welsh Development Agency
Hamilton G.	Manager, Delabole quarry, RTZ Mining and Exploration plc
Howett K.	Mountaineering Council for Scotland
Hughes B.	Commercial Manager, Alfred McAlpine Slate Products Ltd, Penrhyn Quarry, Bethesda

Hughes D.	Planning Department, Arfon Borough Council	Richards I.	Richards, Moorehead and Laing Ltd (formerly Director, Robinson, Jones Partnership Ltd)
Hughes T.	Market Development Manager, Alfred McAlpine Slate Products Ltd	Roberts D.	Keeper, Museum of the North, National Museum of Wales
Jones M.	Chief Planning Officer, Cornwall County Council	Roberts T.	Furness and Cartmel Tourism
Law C.	Director, Alfred McAlpine Slate Products Ltd	Roberts W.	Manager, Gloddfa Ganol Slate Quarry, Blaenau Ffestiniog
Lawday R.	Land Reclamation Department, Welsh Development Agency	RSPB	Royal Society for the Protection of Birds, Dolgellau
MacDonald I.	Scottish Natural Heritage	Sparkes R.	Building Design Partnership, Glasgow
NSQA	Natural Slate Quarries Association, London	Thorpe M.	Environmental Health Department, Meirionnydd District Council
Palmer J.	Richards, Moorehead and Laing Ltd	Wakelin P.	Cadw: Welsh Historic Monuments, Cardiff
Quarry Managers	The managers of various quarries provided this information	Williams M.	Industrial Archaeology Consultant
Rendell T.	Planning Department, Gwynedd County Council		

Printed in the United Kingdom for HMSO
Dd 300050 C7 5/95 13110